Optical Properties of Solids

PATHWAY TO SINGLE-PHOTON SOURCE

Yu-Tsung Tsai

innovativeink
PUBLISHING

Cover image Shutterstock.com

innovativeink PUBLISHING

www.innovativeinkpublishing.com
Send all inquiries to:
4050 Westmark Drive
Dubuque, IA 52004-1840

Print ISBN: 979-8-3197-1222-6
eBook ISBN: 979-8-3197-1223-3

Published in the United States of America

Table of Contents

I

Basic Reviews

CHAPTER 1

Background of Light–Matter Interactions

Figure 1.1 Red apples under sunlight.

It all starts with a simple childish question: why is the apple red, as shown in Figure 1.1? If you use any AI help, the answer is as follows: "An apple appears red because pigments in its skin, mainly anthocyanins, absorb most wavelengths of white light except for those in the red region of the spectrum. When light shines on the apple, the blue and green wavelengths are absorbed, while the red wavelengths are reflected. Since the color we perceive is the light that reaches our eyes after reflection, the dominance of reflected red light makes the apple look red." Even with this simplified statement, one equation should be discussed:

$$I(\lambda) = I_0(\lambda)\, e^{-\alpha(\lambda)d} \tag{1.1}$$

where $I_0(\lambda)$ is incident spectral irradiance, $I(\lambda)$ is transmitted/reflected spectral irradiance, $\alpha(\lambda)$ is the wavelength-dependent absorption coefficient, and d is the path length. Moreover, for this optical experiment, four fundamental components are involved: (a) the light source, sunlight; (b) the optical path, air; (c) the sample, an apple; and (d) the detector, the human eye. Essentially, there is no difference between this simple observation and a standard lab experiment illustrated in Figure 1.2. Although the main purpose of this book is to introduce optical properties in solids, it is beneficial for us to learn these critical components. First, the light source, or more broadly speaking, the electromagnetic wave source, can be categorized into short-wavelength waves, such as gamma rays, all the way to long-wavelength waves, such as radio waves, as sketched in Figure 1.3. However, based on standard laboratory environments, we will focus on the range from extreme ultraviolet (EUV) light above 10 nm to far-infrared light below 1 mm, depicted in Figure 1.4. These are the available light sources, for example, lasers, in the lab (by the way, what does laser stand for?). Second, as shown in Figure 1.2, there will often be some optical components such as filters, mirrors, or objectives along the optical path. As you can imagine, these components are all "solids," and the light–matter interaction should also include their responses before the light reaches the sample or specimen. Third, the light will interact with the sample, and lastly, the transmitted or, in this case, reflected light out of the sample will be guided to the detector for us to analyze. Therefore, to properly study the "optical properties of solids," a few basic principles should be reviewed: the description of electromagnetic waves to understand the behavior of the light source; the description of electronic structures in solids to distinguish various behaviors among insulators, metals, or even superconductors; and the description of the advances of modern physics, specifically quantum mechanics, to utilize perturbation theory, oscillator strength, and quantized photons.

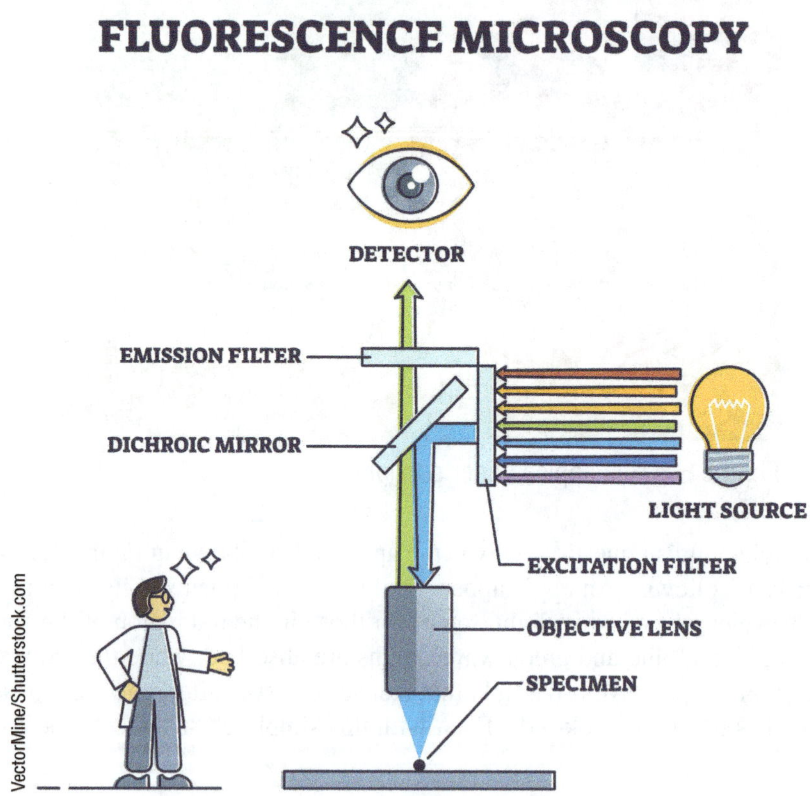

Figure 1.2 Optic experiment in a typical laboratory.

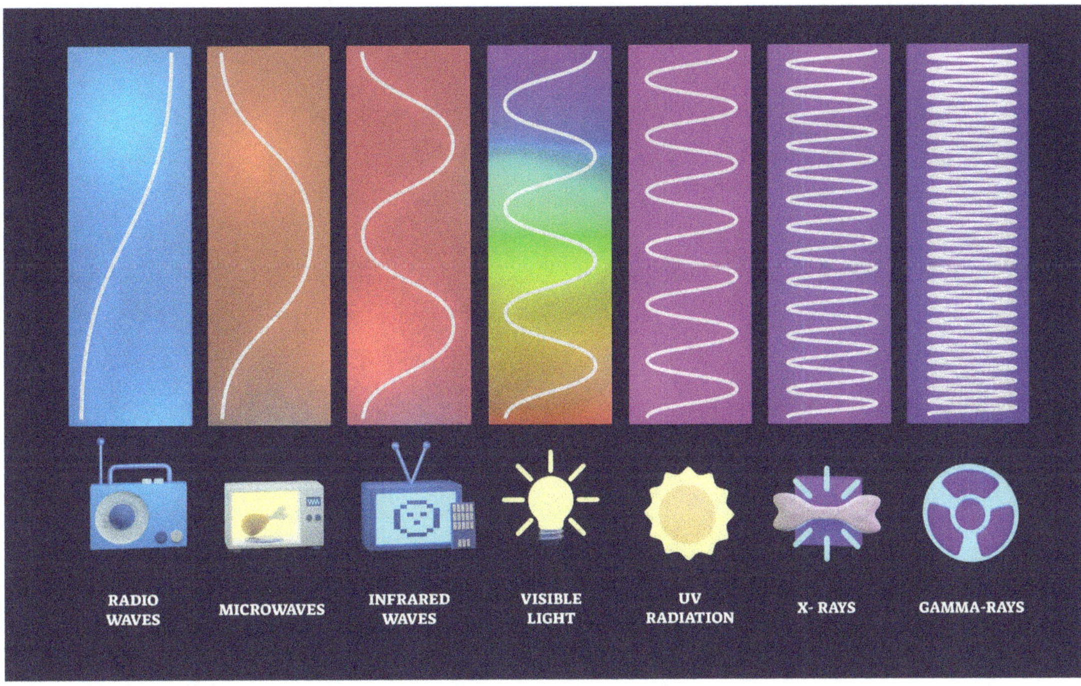

Figure 1.3 Wave source categorization.

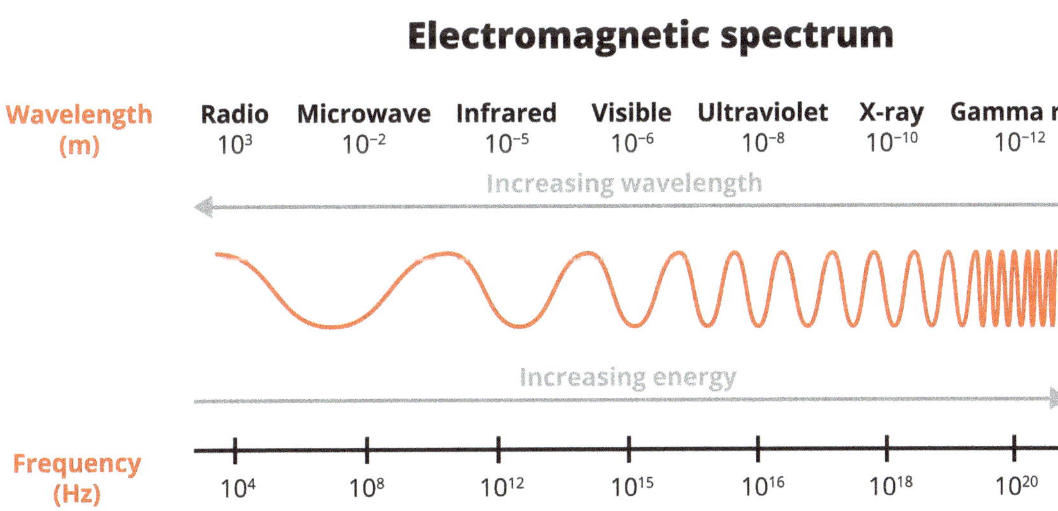

Figure 1.4 Electromagnetic spectrum with corresponding wavelength and frequency.

CHAPTER 2

Review of Maxwell Equations and Plane Waves

2.1 CAUCHY INTEGRAL IN COMPLEX PLANE

Before we dive into the description of electromagnetic (EM) waves, we need to review a mathematical method of complex analysis, the Cauchy integral. The Cauchy integral in the complex plane is important for EM waves because many key wave propagation quantities are analytic functions of complex frequency or wave number and therefore obey the Cauchy integral formula. Because physical EM systems must satisfy causality (a material cannot respond before it is stimulated), their response functions are analytic in the upper half of the complex plane, allowing contour integration to be used to evaluate fields, identify resonances, and derive relationships such as the Kramers–Kronig relations that link absorption to dispersion. Through contour deformation and the residue theorem, we can then compute wave propagation, decay, and mode structure in ways that would be impossible on the real axis alone. This analyticity allows the use of the Cauchy integral formula,

$$f(Z_0) = \frac{1}{2\pi i} \oint_C \frac{f(z)}{z - z_0} dz \tag{2.1}$$

to relate values of EM response functions at different frequencies and to evaluate integrals that determine wave propagation. Contour integration and the residue theorem, shown in Figure 2.1, also allow the evaluation of Fourier integrals used to compute fields, such as

$$E(t) = \frac{1}{2\pi} \int_{-\infty}^{\infty} E(\omega) e^{-i\omega t} d\omega \tag{2.2}$$

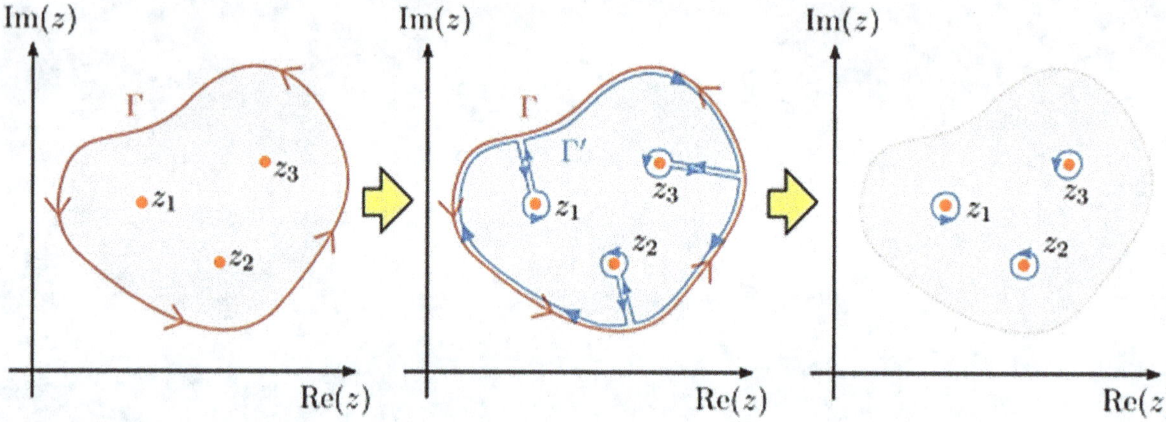

Figure 2.1 Cauchy integral with three residues.

Source: Y. D. Chang, LibreTexts Physics.

by enclosing poles corresponding to resonances and guided modes. The residue theorem is a fundamental principle in complex analysis, which states that the line integral of an analytic function, $f(z)$, around a simple closed contour, C, is equal to $2\pi i$ times the sum of the residues of $f(z)$ at its isolated singularities, z_k, that lie inside the contour, as expressed in Eq. 2.5 and Figure 2.1. This theorem is exceptionally powerful because it allows complex integrals, which are often computationally difficult, to be evaluated simply by identifying the singularities enclosed by the contour and calculating the corresponding residues, a calculation that varies depending on whether the singularity is a simple pole (Eq. 2.3) or a pole of higher order ***m*** (Eq. 2.4).

$$Res_{z=z_0} f(z) = \lim_{z \to z_0} (z - z_0) f(z) \tag{2.3}$$

$$Res_{z=z_0} f(z) = \frac{1}{(m-1)!} \lim_{z \to z_0} \frac{d^{m-1}}{dz^{m-1}} \left((z - z_0)^m f(z) \right) \tag{2.4}$$

$$\oint_C f(z) dz = 2\pi i \sum_{k=1}^{n} Res(f, z_k) \tag{2.5}$$

You can already see that these singularities correspond to the resonance frequency that is critical in describing EM waves. More examples will be introduced in later chapters.

2.2 MAXWELL'S EQUATIONS

Maxwell's equations are a set of four fundamental laws in classical physics that comprehensively describe the behavior of electric $\boldsymbol{E}$ and magnetic $\boldsymbol{B}$ fields and their relationship with electric charges and currents. These laws are as follows: Gauss's law for electricity, which states that electric fields originate from electric charges; Gauss's law for magnetism, which states that magnetic fields have no sources or sinks, meaning magnetic field lines are always closed loops; Faraday's law of induction, which describes how a changing magnetic field creates an electric field; and the Ampère–Maxwell law, which states that magnetic fields are created by electric currents and, crucially, by a changing electric field (Maxwell's displacement current term, as shown in Eq. 2.6). Together, these equations unify electricity and magnetism into a single EM force and predict the existence of EM waves, which travel at the speed of light.

$$\nabla \cdot \vec{E} = \rho$$

$$\nabla \cdot \vec{B} = 0$$

$$\nabla \times \vec{E} = -\partial \vec{B} / \partial t$$

$$\nabla \times \vec{B} = \mu_0 \varepsilon_0 \partial \vec{E} / \partial t + \mu_0 \vec{J} \tag{2.6}$$

2.3 PLANE WAVES

Now, we usually consider light as EM waves and follow plane wave forms. Why? Light is understood as a plane wave primarily because it is the simplest and most fundamental traveling wave solution derived from Maxwell's equations in a vacuum. These equations require the electric ***E*** and magnetic ***B*** fields to oscillate mutually perpendicularly and to be transverse to the direction of propagation, which defines the mathematical structure of a plane wave. Physically, while light originates as a spherical wave from any source, the far-field approximation dictates that when the waves are observed at a great distance, the local curvature of the wave fronts becomes negligible, causing the small observable portion of the wave to appear as a flat, parallel, and uniform plane (what is the definition of far field?). The plane wave solutions in a vacuum are described here and illustrated in Figure 2.2:

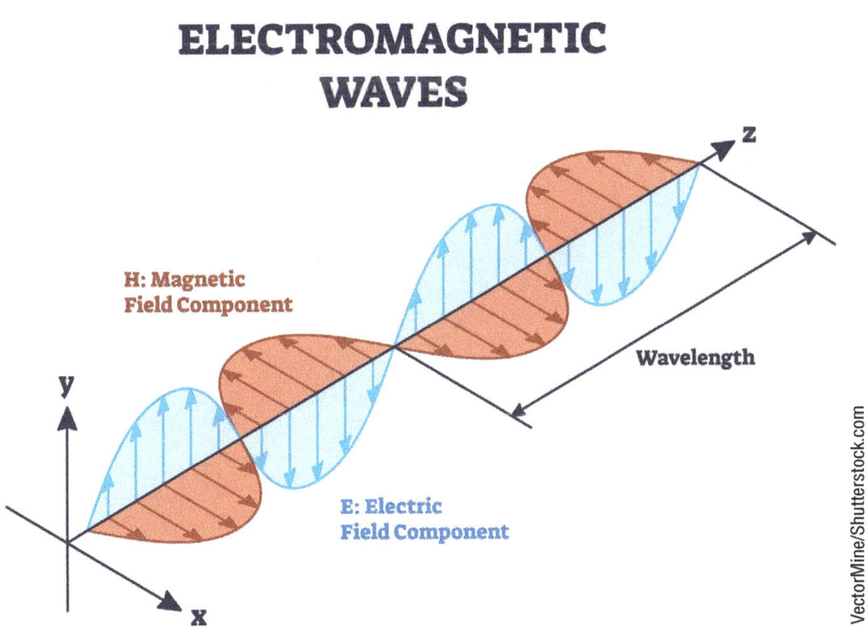

Figure 2.2 Plane wave propagation showing electric field and magnetic field oscillating perpendicular to each other.

$$\tilde{E}(z,t) = \tilde{E}_0 e^{i(kz-\omega t)}$$

$$\tilde{B}(z,t) = \tilde{B}_0 e^{i(kz-\omega t)} \tag{2.7}$$

We will use these forms in the following chapters to describe light propagation.

2.4 PLANE ELECTROMAGNETIC WAVES IN SOLIDS

The relationship between the macroscopic fields (*D* and *H*) and the primary fields (*E* and *B*) in a medium is established through the constitutive relations. Specifically, for a linear dielectric, the electric displacement is linked to the electric field by $D = \mu E$, and the magnetic flux density is linked to the magnetic field by $B = \mu H$. These relations encapsulate the material's properties within the permittivity (ϵ) and permeability (μ) of the medium, demonstrating precisely how the fields are altered due to the polarization and magnetization induced within the material:

$$\nabla \cdot \vec{D} = 0$$

$$\nabla \cdot \vec{B} = 0$$

$$\nabla \times \vec{E} = -\frac{1}{c}\frac{\partial \vec{B}}{\partial t}$$

$$\nabla \times \vec{H} = -\frac{1}{c}\frac{\partial \vec{D}}{\partial t} + \frac{4\pi}{c}\vec{J} \tag{2.8}$$

As a result, the plane wave solution in Eq. 2.7 only needs to be modified with permittivity (ϵ) and permeability (μ), replacing permittivity and permeability in vacuum.

CHAPTER 3

Review of Drude Model and Lorentz Model in Solids

3.1 DRUDE MODEL FOR METALS

The Drude model is a classical theory that describes how electrons behave in metals and how this behavior determines their electrical and optical properties. This model treats conduction electrons as a gas of free particles that move randomly and collide with ions in the lattice. These collisions occur with an average time between them called the relaxation time, τ. When an electric field $\boldsymbol{E}$ is applied, electrons accelerate but are continually slowed by these collisions, leading to the equation of motion:

$$m\frac{d\mathbf{v}}{dt} = -e\mathbf{E} - \frac{m\mathrm{v}}{\tau} \tag{3.1}$$

Solving this equation gives the Drude electrical conductivity:

$$\sigma_{DC} = \frac{ne^2\tau}{m} \tag{3.2}$$

where n is the electron density.

Since our focus is on light–matter interaction, in this case, light–charge carrier interactions in metals, we will utilize the Drude model and learn how it describes metals responding to electromagnetic waves by giving a frequency-dependent (complex) conductivity derived in Eq. 3.6 (instead of the conductivity σ_{DC} in Eq. 3.2) and the complex dielectric function in Eq. 3.11.

With the applied electric field,

$$E_{ext}(z,t) = E_0 \tag{3.3}$$

And considering damping (collision from the Drude model),

$$m\ddot{r} = -eE_{ext} - \frac{m}{\tau}\dot{r} \tag{3.4}$$

$$\dot{r} = \frac{-e}{m\left(1/\tau - i\omega\right)} \tag{3.5}$$

$$\sigma\left(\omega\right) = \frac{ne^2\tau / m}{1 - i\omega\tau} \tag{3.6}$$

However, the real part and imaginary parts are in fact

$$\sigma_R\left(\omega\right) = \frac{\sigma_{DC}}{1 + \omega^2\tau^2} \tag{3.7}$$

$$\sigma_i\left(\omega\right) = \frac{\omega\tau\sigma_{DC}}{1 + \omega^2\tau^2} \tag{3.8}$$

And the plasma frequency is defined as follows:

$$\omega_{pl} = \sqrt{4\pi ne^2\tau / m} \tag{3.9}$$

so that the conductivity can be rewritten as follows:

$$\sigma\left(\omega\right) = \frac{\omega_{pl}{}^2 / 4\pi}{1 - i\omega\tau} \tag{3.10}$$

However, the *Drude dielectric function* is

$$\varepsilon\left(\omega\right) = \varepsilon_\infty - \frac{\omega_{pl}^2}{\omega\left(\omega + \frac{i}{\tau}\right)} \tag{3.11}$$

Splitting $\varepsilon(\omega)$ gives

$$\varepsilon\left(\omega\right) = \varepsilon_\infty - \frac{\omega_p^2\left(\omega^2 - \frac{i\omega}{\tau}\right)}{\omega^2 + (1/\tau)^2} \tag{3.12}$$

with the real part (dispersion):

$$\varepsilon_1(\omega) = \varepsilon_\infty - \frac{\omega_p^2}{\omega^2 + (1/\tau)^2} \tag{3.13}$$

For most metals, $\omega_p \gg \frac{1}{\tau}$; therefore,

$$\varepsilon_1(\omega) \approx \varepsilon_\infty - \frac{\omega_p^2}{\omega^2} \tag{3.14}$$

and the imaginary part (absorption):

$$\varepsilon_2(\omega) = \frac{\frac{\omega_p^2}{\tau}}{\omega\left[\omega^2 + (1/\tau)^2\right]} \tag{3.15}$$

This sets a boundary: If $\omega < \omega_p$, the metal reflects strongly. On the other hand, if $\omega > \omega_p$, the metal becomes transparent. That is because a negative dielectric function means the wave vector inside the metal becomes imaginary:

$$k = \frac{\omega}{c}\sqrt{\varepsilon(\omega)} \tag{3.16}$$

So the electromagnetic wave cannot propagate and instead decays exponentially (skin effect), and hence, the light is reflected.

For frequencies above the plasma frequency, electrons are too heavy (inertia-dominated) to keep up with the rapid oscillations of the field. They can no longer screen it effectively. In this regime, $\varepsilon_1(\omega) > 0$, so the wave vector in Eq. 3.16 becomes real, meaning the electromagnetic wave can propagate inside the metal. When waves propagate instead of being screened out, the medium becomes transparent. The *plasma frequency* is the characteristic frequency above which metals become transparent and below which they reflect light. This explains why metals are shiny in the visible range and why they support surface plasmons.

3.2 LORENTZ MODEL FOR INSULATORS

The Drude model, which treats electrons as free particles, is unsuitable for describing electron behavior in insulators because their electrons are not free to move, as sketched in Figure 3.1. Instead, the Lorentz model is employed to describe these bound electrons. This model describes how an electron within an insulator responds to an oscillating electromagnetic field by treating the electron as a driven and damped

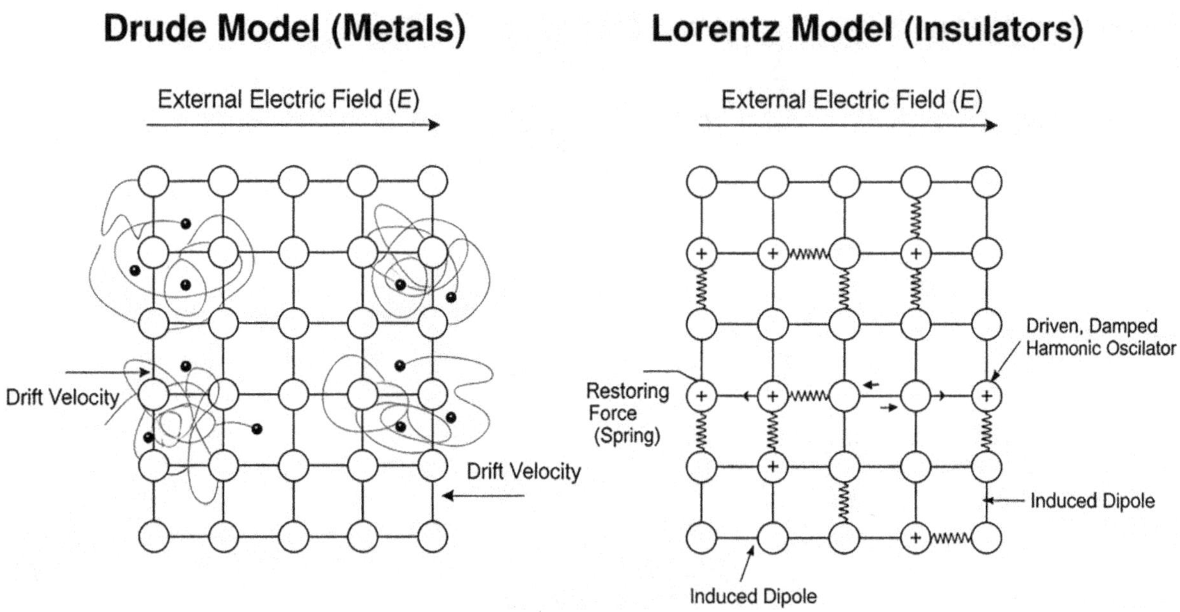

Diagram generated by AI assistance.

Figure 3.1 Comparison of Drude model and Lorentz model.

harmonic oscillator attached to its equilibrium position by a restoring spring-like force. This approach allows for the calculation of the material's polarization and ultimately its permittivity. When an electric field $Ee^{-i\omega t}$ drives the electron, its displacement $x(t)$ satisfies the oscillator equation with harmonic potential:

$$m\ddot{\boldsymbol{r}} = \boldsymbol{F}_{harmonic} + \boldsymbol{F}_{damping} + \boldsymbol{F}_{external}$$

$$m\ddot{\boldsymbol{r}} = -m\omega_0^2\boldsymbol{r} - m\gamma\dot{\boldsymbol{r}} - e\boldsymbol{E}_0 e^{-i\omega t}$$

$$\boldsymbol{r} = \frac{-e / m}{\omega_0^2 - \omega^2 - i\omega\gamma}\boldsymbol{E} \tag{3.17}$$

We can picture this strength as polarizability by considering the dipole moment:

$$\boldsymbol{p} = -e\boldsymbol{r} = \alpha_p \boldsymbol{E} \tag{3.18}$$

and define the *polarizability*, which tells you how easily the electron cloud can be distorted by the electric field:

$$\alpha_p = \frac{e^2 / m}{\omega_0^2 - \omega^2 - i\omega\gamma} \tag{3.19}$$

where ω_0 is the natural resonance frequency of the bound electron and γ is the damping constant. Solving this leads to the Lorentz dielectric function,

$$\varepsilon(\omega) = \varepsilon_\infty + \frac{f\omega_p^2}{\omega_0^2 - \omega^2 - i\gamma\omega} \tag{3.20}$$

where $\omega_p^2 = ne^2/(\varepsilon_0 m)$ is the effective oscillator plasma frequency and f is the oscillator strength. Far from resonance ($\omega \neq \omega_0$), bound electrons respond weakly, and absorption is small, making the insulator transparent; near resonance, strong polarization, and damping enhance $\varepsilon_2(\omega)$, causing strong absorption and reducing transmission. This model explains why insulators exhibit strong absorption near $\omega \approx \omega_0$, why their refractive index varies with frequency (dispersion), and why their dielectric response is purely due to bound-electron oscillations rather than free carriers.

3.3 BAND STRUCTURE ANALYSIS FOR SEMICONDUCTORS

One critical category of materials in our modern life is semiconductors. A semiconductor is a material that has an electrical conductivity value falling between that of a conductor (like copper) and an insulator (like glass). This unique property arises from its electronic band structure: at absolute zero temperature, its valence band (VB) is full, and its conduction band (CB) is empty, separated by a relatively small bandgap (typically 1–3 eV). At room temperature, thermal energy allows a small number of electrons to jump this gap, enabling some current flow, and this conductivity can be precisely and controllably increased by adding minute amounts of impurities in a process called doping. Semiconductors, particularly silicon (Si) and gallium arsenide (GaAs), are critically important because their controllable conductivity makes them the essential foundation for virtually all modern electronic devices, including transistors, diodes, integrated circuits (microchips), solar cells, and LEDs, enabling the storage, processing, and transmission of digital information that powers contemporary technology.

The light–semiconductor interaction is fundamentally governed by the interband transition, where a photon's energy dictates whether it can be absorbed or emitted by the material.

Interband transitions are most important near the band extrema—the top of the VB and the bottom of the CB. This is primarily because the bandgap represents the minimum energy threshold for any transition, which occurs only between states right at these extrema. Furthermore, due to the Fermi–Dirac distribution, the highest population of electrons available for excitation is found near E_v and the most readily available empty states are located near E_c. Consequently, transitions involving these states dominate the overall optical and electronic response of the semiconductor, simplifying initial theoretical analysis by allowing the use of the effective mass approximation (parabolic bands) for the nearby energy states.

For absorption to occur, an electron in the fully occupied VB must absorb a photon and gain enough energy to jump across the forbidden bandgap (E_g) into the empty CB, thereby creating an electron–hole pair. The essential condition for this process is energy conservation:

$$h\upsilon \geq E_g$$

This results in a change in the material's conductivity.

The following equation describes the interband conductivity tensor ($\sigma_{\alpha\beta}$) of a semiconductor, which links the current density ($j\alpha$) in direction α to the applied electric field (E_β) oscillating at frequency ω.

$$j_\alpha = \sigma_{\alpha\beta} E_\beta$$

$$\sigma_{\alpha\beta} = -\frac{e^2}{m^2} \sum_{i,j} \left[\frac{f(E_i) - f(E_j)}{E_i - E_j} \right] \frac{\langle i|p_\alpha|j\rangle \langle i|p_\beta|j\rangle}{\left[-i\omega + \tau^{-1} + (i/\hbar)\left(E_i - E_j\right) \right]} \tag{3.21}$$

This tensor is derived from quantum mechanics and the Fermi's golden rule, describing the rate at which electrons transition from the occupied VB to the unoccupied CB (interband transitions). The tensor's value depends critically on three factors: the population difference between the initial and final states (governed by the Fermi–Dirac distribution); the *momentum matrix elements* $\left(\langle i|p|j\rangle\right)$ which enforce momentum conservation and determine the transition strength; and the *resonance term*.

$$\frac{1}{\left[-i\omega + \tau^{-1} + (i/\hbar)\left(E_i - E_j\right) \right]}$$

which shows that maximum conductivity (absorption) occurs when the photon energy is *resonant with the energy difference*, accounting for damping due to scattering. This complex formula models the frequency-dependent optical and electrical response of the material.

Moreover, the direct or indirect transition of semiconductors determines the efficiency of light interaction, a difference driven by the strict need for crystal momentum conservation. In direct bandgap materials (like GaAs), the VB maximum and CB minimum align in momentum space, allowing the electron to absorb a photon and jump straight across the gap vertically. This two-particle interaction (electron + photon) is highly probable, leading to efficient light absorption and emission. In contrast, indirect bandgap materials (like Si) have misaligned extrema, requiring a third particle, a phonon (lattice vibration quantum), to carry the necessary momentum difference, making the three-particle interaction significantly less probable and thus resulting in weak light emission.

$$\vec{k}_v = \vec{k}_c \pm \vec{q}$$

Also, the actual rate and strength of the coupling between the electromagnetic wave and the electrons in the semiconductor are quantified by the Fermi's golden rule, which is derived from time-dependent perturbation theory. This rule treats the light as a weak perturbation acting on the electron states and states that the transition probability rate is proportional to the square of the momentum matrix element. This matrix element is crucial as it embeds the selection rules—including momentum conservation—that govern which specific initial state in the VB can transition to a final state in the CB, thereby quantifying the material's absorption coefficient.

3.4 JOINT DENSITY OF STATES AND CRITICAL POINTS

The joint density of states (JDOS) is a crucial function in solid-state physics that dictates a semiconductor's optical behavior. It quantifies the number of available pairs of electronic states—one state in the occupied VB and one in the unoccupied CB—that are separated by an energy exactly equal to the incident photon energy. The JDOS is essential because the optical absorption rate for light is directly proportional to it, where the term involving the momentum operator $\boldsymbol{P}$ dictates the transition probability.

The calculated shape of the JDOS is governed by the energy difference between the bands as a function of crystal momentum, and its most distinct features are its critical points. A critical point occurs at any point in the Brillouin zone, where the gradient of the energy separation between the bands vanishes. This vanishing gradient causes the denominator in the JDOS integral to become zero, resulting in a sudden, singular change in the JDOS function. Physically, this signifies that the energy bands are locally parallel in $\boldsymbol{k}$ space, creating a very high density of states available for photo excitation at that precise energy.

These critical points are the origins of the sharp onset and features observed in the material's optical absorption spectrum. For instance, the absolute minimum of the bandgap corresponds to the primary critical point, marking the minimum energy threshold for interband absorption. Higher-energy critical points correspond to higher-energy features in the ultraviolet (UV) absorption spectrum and are vital fingerprints used to map and characterize the entire electronic band structure of the semiconductor. In short, critical points are k points where both slopes in the CB or VB are 0 or identical.

JDOS means the combination of density of states at both the CB and VB:

$$D_{CV}\left(\hbar\omega\right)=\frac{2}{8\pi^3}\int\delta\left[E_c\left(\boldsymbol{k}\right)-E_v\left(\boldsymbol{k}\right)-\hbar\omega\right]d^3\boldsymbol{k}$$

$$D_{CV}\left(\hbar\omega\right)=\frac{2}{8\pi^3}\iint\frac{d\boldsymbol{S}}{\left|\nabla_k\left(E_c-E_v\right)\right|_{E_c-E_v=\hbar\omega}} \tag{3.22}$$

At some points where E_c–E_v is stationary, called *joint critical points*, also known as Van Hove singularities, the denominator vanishes, usually at high symmetry points in the Brillouin zone.

The JDOS is

$$D_{CV}\left(\hbar\omega\right)=\left[\frac{m_r k}{\pi^2\hbar^2}\right]_{E_c-E_v=\hbar\omega} \tag{3.23}$$

with

$$\frac{1}{m_r}=\frac{1}{m_v}+\frac{1}{m_c} \tag{3.24}$$

CHAPTER 4

Review of Quantum Mechanics: Optical Excitations

4.1 PERTURBATION EXPANSION

By the nature of light–matter interaction, when an electromagnetic (EM) wave interacts with charge carriers, the response is time-dependent. Also, due to advances in solid-state band structure calculations and modern physics, discrete density of states for carriers and quantized light (photons) are widely implemented to describe carrier behaviors. Therefore, it is necessary to revisit the critical time-dependent Schrödinger equation:

$$i\hbar\frac{\partial\psi}{\partial t}=\left(\mathrm{H}_0+\mathrm{H}'\right)\psi \tag{4.1}$$

where the unperturbed Hamiltonian gives stationary (time-independent) $\mathrm{H}_0\varphi_n=E_n\varphi_n$, with eigenfunctions φ_n and eigenvalues E_n.

And H′(*t*) is the time-dependent perturbation with wave function

$$\psi\left(\vec{r},t\right)=\sum_n a_n\left(t\right)u_n\left(\vec{r}\right)e^{-iE_nt/\hbar} \tag{4.2}$$

The time-dependent solution with expansion, first-order perturbation theory gives

$$\dot{a}_k\left(t\right)=\frac{1}{i\hbar}\sum\nolimits_n a_n\left(t\right)e^{i\omega_{kn}t}\left\langle k|\mathrm{H}'\left(t\right)|n\right\rangle \tag{4.3}$$

$$\left|a_m^{(1)}\left(t\right)\right|^2=\left(\frac{\left|\langle m|\mathrm{H}'|l\rangle\right|^2}{\hbar^2}\right)\left(\frac{4sin^2\left(\omega_{ml}t/2\right)}{\omega_{ml}^2}\right) \tag{4.4}$$

Here, let us revisit the definition of perturbation theory in quantum mechanics. In quantum mechanics, perturbation theory is an approximation method used when the Hamiltonian of a system differs only slightly from one that can be solved exactly. The total Hamiltonian is written as follows:

$$H = H_0 + \lambda V \tag{4.5}$$

where H_0 is the exactly solvable Hamiltonian, V is a small perturbing potential, and λ is a parameter indicating the perturbation's strength (set to 1 at the end), sketched in Figure 4.1.

This perturbation causes the original, exact energy levels $E_n(0)$ and states $\mu_n\,(0)$ of the simple system to shift to new, approximate values (E_n and μ_n). Perturbation theory calculates these new values by treating the shift as a series of corrections. The resulting energy levels, denoted as E'_n, are shown to be slightly displaced from the original lines. The eigenvalues and eigenstates of H are then expressed as power series corrections to those of H_0, for example, the energy expansion:

$$E_n = E_n^{(0)} + \lambda E_n^{(1)} + \lambda^2 E_n^{(2)} + \cdots \tag{4.6}$$

In time-dependent perturbation theory, the perturbation $V(t)$ varies with time and induces transitions between the unperturbed eigenstates of H_0, as shown in Figure 4.2.

Figure 4.2 shows a comparison between an unperturbed quantum system and the same system after a perturbation is applied. On the left side, the diagram shows a single energy level with its corresponding unperturbed wave function—smooth and regular, representing the exact solution of the original Hamiltonian. With a perturbing potential $V(t)$, the energy level is shifted slightly upward or downward, illustrating that the new energy becomes $E_n = E_n^0 + \Delta E_{small}$. Below this shifted level, the wave function is noticeably altered, showing distortions compared to its original shape. Essentially, a small additional potential modifies both the energies and the shapes of the wave functions, but only slightly, allowing the new solutions to be expressed as corrections to the known unperturbed ones.

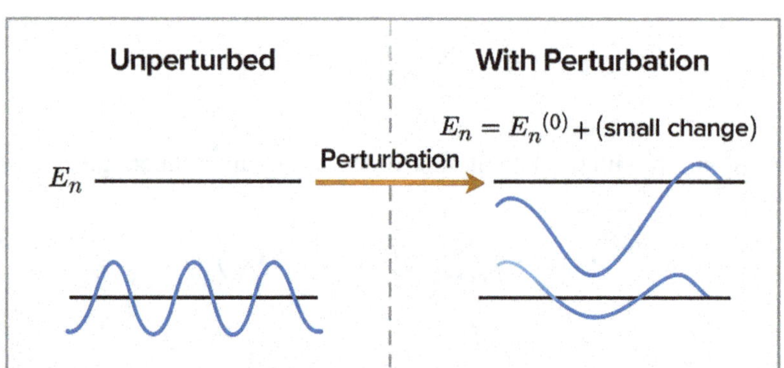

Sketch generated with AI assistance.

Figure 4.1 Energy changed due to the small perturbation, as well as the wave function affected by this potential.

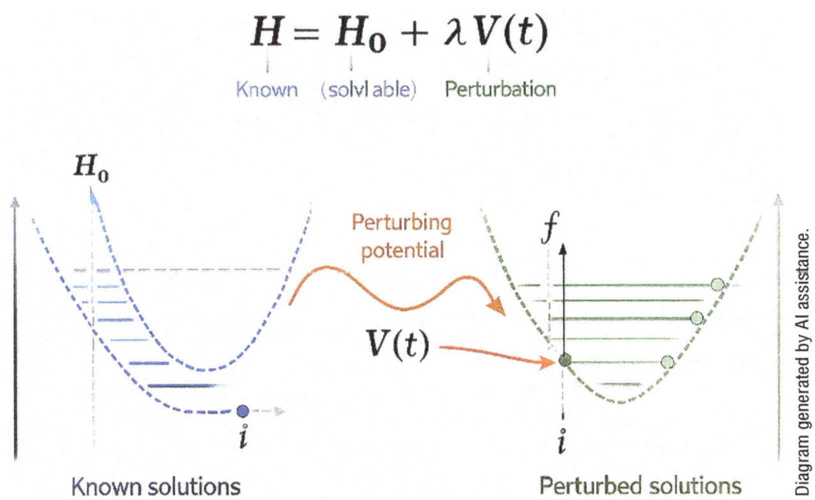

Figure 4.2 A comparison of a quantum system before and after a small perturbation is applied. The unperturbed state has well-defined energy levels and wave functions, while the perturbed state shows slightly shifted energies and modified wave function shapes, illustrating the fundamental idea of perturbation theory.

Consider the following example: an atom exposed to a weak oscillating electric field:

$$V(t) = -eE_0 x\cos(\omega t) \tag{4.7}$$

This perturbation can drive transitions between atomic energy levels $|i\rangle \rightarrow |f\rangle$. The transition probability amplitude is proportional to the dipole matrix element $\langle f|x|i\rangle$ and is resonantly enhanced when the field frequency matches the energy gap, $\hbar\omega \approx E_f - E_i$.

The first-order wave function is as follows:

$$|\Psi(t)\rangle \approx e^{-\frac{iE_i^{(0)}t}{\hbar}}|i\rangle + \sum\nolimits_{f\neq i} a_f^{(1)}(t)e^{-\frac{iE_f^{(0)}t}{\hbar}}|f\rangle \tag{4.8}$$

with the following two important cases:

a) Near resonance $\omega \approx \omega_{fi}$: the second denominator is small and dominates, so

$$a_f^{(1)}(t) \approx \frac{eE_0 x_{fi}}{2\hbar}\frac{e^{i(\omega_{fi}-\omega)t}-1}{\omega_{fi}-\omega} \tag{4.9}$$

b) Exact resonance $\omega = \omega_{fi}$: with short times (first order still valid and phase details dropped):

$$a_f^{(1)}(t) \xrightarrow{\omega=\omega_{fi}} \frac{eE_0 x_{fi}}{2\hbar}t \tag{4.10}$$

This describes photon absorption leading to the excitation of an electron to a higher atomic state.

4.2 FERMI'S GOLDEN RULE

Following the time-dependent perturbation theory, I would like to revisit the Heisenberg energy–time uncertainty principle. It states that a system observed over a finite time interval Δt has an intrinsic uncertainty in energy:

$$\Delta E \Delta t \gtrsim \frac{\hbar}{2}$$

If a system is driven by a perturbation for only a finite time t, then the transitions it produces cannot be sharp at a single energy; instead, the energy spectrum has a width of order:

$$\Delta E \sim \frac{\hbar}{t}$$

This idea foreshadows that transitions from an initial energy E_i to a final energy E_f will only be sharply energy-conserving in the limit of long times. The *Fermi's golden rule* emerges exactly from this long-time limit.

In first-order time-dependent perturbation theory, the transition amplitude from i to f after time t is

$$c_f(t) = -\frac{i}{\hbar}\int_0^t \langle f|V(t')|i\rangle e^{i\omega_{fi}t'}\,dt' \tag{4.11}$$

with

$$\omega_{fi} = \frac{E_f - E_i}{\hbar} \tag{4.12}$$

If the perturbation is time-independent (or switched on suddenly), $V(t) = V$, then:

$$c_f(t) = -\frac{i}{\hbar}V_{fi}\int_0^t e^{i\omega_{fi}t'}\,dt' \tag{4.13}$$

Evaluate the integral:

$$\int_0^t e^{i\omega t'}\,dt' = t\ \mathrm{sinc}\left(\frac{\omega t}{2}\right)e^{\frac{i\omega t}{2}} \tag{4.14}$$

where

$$\mathrm{sinc}(x) = \frac{\sin x}{x}$$

Thus,

$$\left|c_f(t)\right|^2 = \frac{\left|V_{fi}\right|^2}{\hbar^2}\left|\frac{\sin\left(\frac{\omega_{fi}t}{2}\right)}{\frac{\omega_{fi}}{2}}\right|^2 \tag{4.15}$$

Define the transition probability:

$$P_{i\to f}(t) = \left|c_f(t)\right|^2 \tag{4.16}$$

Use the limit:

$$\lim_{t\to\infty}\left(\frac{\sin(\omega t/2)}{\pi\omega/2}\right)^2 = t\delta(\omega) \tag{4.17}$$

$$P_{i\to f}(t) \approx \frac{\left|V_{fi}\right|^2}{\hbar^2} 2\pi t\delta(\omega_{fi}) = \frac{2\pi}{\hbar}\left|V_{fi}\right|^2 t\delta(E_f - E_i) \tag{4.18}$$

Differentiate with respect to time to get a *transition rate* $W_{i\to f}$:

$$W_{i\to f} = \frac{dP}{dt} = \frac{2\pi}{\hbar}\left|V_{fi}\right|^2 \delta(E_f - E_i) \tag{4.19}$$

This is Fermi's golden rule for a single discrete final state.

Usually, final states form a continuum. Replace

$$\delta(E_f - E_i) \to \rho(E_f)$$

where $\rho(E)$ is the density of final states. Fermi's golden rule becomes

$$W_{i\to f} = \frac{2\pi}{\hbar}\left|V_{fi}\right|^2 \rho(E_f) \tag{4.20}$$

Therefore, the transition rate is proportional to the square of the matrix element and also proportional to the density of accessible final states at the same energy. Exact energy conservation emerges only in the long-time limit—a direct consequence of the energy—time uncertainty principle.

$$P_m = \frac{1}{t}\sum_{m' \approx m} \left| a_{m'}^{(1)}(t) \right|^2 \tag{4.21}$$

$$P_m = \frac{1}{\hbar^2 t}\int \left| 4\mathrm{H}'_{ml} \right|^2 \left(\frac{sin^2(\omega_{m'l} t/2)}{\omega_{m'l}^2} \right) \rho(E_{m'})\, dE_{m'} \tag{4.22}$$

Prove Fermi's golden rule:

$$P_m \cong (2\pi/\hbar) \left| H'_{ml} \right|^2 \rho(E_m) \tag{4.23}$$

4.3 THE OSCILLATOR STRENGTH

Oscillator strength is a dimensionless measure of how strongly an electronic transition in an atom, molecule, or solid interacts with EM radiation. It plays a central role in quantum mechanics and optical spectroscopy because it links the quantum description of transitions to observable absorption or emission intensities.

In quantum mechanics, the oscillator strength for a transition from an initial state $|i\rangle$ to a final state $|f\rangle$ is defined in terms of the electric-dipole matrix element. For an electronic transition with energy difference $\Delta E = E_f - E_i$, the (dipole) oscillator strength is

$$f_{if} = \frac{2m_e}{3\hbar^2}(E_f - E_i)\left| \langle f | \mathbf{r} | i \rangle \right|^2 \tag{4.24}$$

where m_e is the electron mass and $\langle f | \mathbf{r} | i \rangle$ is the transition dipole moment. A large dipole matrix element or large energy spacing increases the oscillator strength, indicating a more "radiatively active" transition. When selection rules forbid a transition (e.g., parity or angular momentum restrictions), the dipole matrix element vanishes, and $f_{if} = 0$. The oscillator strength obeys the Thomas–Reiche–Kuhn (TRK) sum rule, which states that the total oscillator strength from a given initial state sums to the number of electrons:

$$\sum_f f_{if} = N_e \tag{4.25}$$

This expresses a fundamental conservation property of charge and ensures that oscillator strengths provide a complete distribution of transition probabilities.

In optics and spectroscopy, oscillator strength determines how strongly a transition absorbs light. This appears in the relationship between oscillator strength and the frequency-dependent absorption cross-section $\sigma(\omega)$. Integrating the cross-section over all frequencies yields

$$\int \sigma(\omega)\, d\omega = \frac{\pi e^2}{2\varepsilon_0 m_e c} f_{if} \tag{4.26}$$

showing that the *area* under an absorption peak is proportional to f_{if}. A higher oscillator strength produces a stronger, more intense absorption line. Likewise, the strength of spontaneous emission from state $|f\rangle$ to $|i\rangle$ is directly connected to oscillator strength through the Einstein A coefficient, which scales approximately as

$$A_{fi} \propto f_{if}\,(E_f - E_i)^2 \tag{4.27}$$

Thus, transitions with larger oscillator strengths have larger emission rates and are more favorable for laser action or strong fluorescence.

From a classical perspective, oscillator strength links quantum transitions to the classical Lorentz model of an electron bound by a restoring force and driven by light. Each quantum transition behaves like such a classical oscillator with a certain effective "weight" equal to its oscillator strength. A transition with $f \approx 1$ acts like a single classical electron oscillator; transitions with $f \ll 1$ are much weaker; and in solids or collective excitations, transitions can have $f > 1$, corresponding to coherent motion of multiple electrons.

In summary, oscillator strength quantifies the intrinsic coupling between a quantum transition and an EM field. It depends on the dipole moment, energy spacing, and electronic structure, and it directly determines the intensity of optical absorption, emission, and many radiative processes.

$$f_{n0} \equiv \frac{2}{m\hbar\omega_{n0}}\left|\langle n|\,p_r\,|0\rangle\right|^2 \tag{4.28}$$

The oscillator strength f_{if} for an electronic transition from an initial state i to a final state f is defined through the dipole matrix element and the transition energy. For an electric-dipole transition, it is given by

$$f_{if} = \frac{2m_e\omega_{if}}{3\hbar e^2}\left|\langle f|\mathbf{r}|i\rangle\right|^2 \tag{4.29}$$

where ω_{if} is the angular frequency of the transition and $\langle f|\mathbf{r}|i\rangle$ is the transition dipole moment. As an example, consider the hydrogen atom transition $1s \rightarrow 2p$. Quantum mechanical evaluation gives a dipole matrix element of $\left|\langle 2p|\mathbf{r}|1s\rangle\right| = 0.744\ a_0$ and a transition energy of 10.2 eV. Substituting these values into the formula above yields a dimensionless oscillator strength of approximately

$$f_{1s\rightarrow 2p} \approx 0.416$$

which reflects a strongly allowed electric-dipole transition responsible for the Lyman-α line.

II

Classical Optical Properties of Solids

CHAPTER 5

Classical Theories for Dielectric Function

5.1 POLARIZATION OF MEDIUMS

The dielectric constant is crucially important to optical properties because it has a direct, mathematical relationship to a material's refractive index n, as expressed by the following formula:

$$n = \sqrt{\varepsilon_r}$$

for nonmagnetic materials at optical frequencies. This fundamental connection means that the dielectric constant effectively determines how light behaves when it encounters a substance; it dictates the speed at which light propagates through the material, which governs how much light bends (refracts) when changing mediums and how much light is absorbed versus transmitted.

Moreover, the dielectric constant determines how a material interacts with electromagnetic (EM) waves by measuring the material's ability to polarize internally when exposed to the wave's oscillating electric field, as sketched in Figure 5.1. This polarization effect directly controls several critical wave behaviors: it dictates the propagation speed and wavelength of the wave within the medium (slower speeds in materials with higher constants); it governs the amount of reflection and transmission that occurs when the wave crosses boundaries between different materials; and the "imaginary" or loss component of the constant determines how much of the wave's energy is absorbed as heat.

5.2 THE COMPLEX DIELECTRIC CONSTANT

The *complex dielectric constant* (or complex permittivity), denoted as $\tilde{\varepsilon}$, is a frequency-dependent material property that describes both how a material stores electrical energy and how it loses that energy when interacting with an EM wave. It is expressed as a complex number with a real part and an imaginary part:

$$\tilde{\varepsilon}(\omega) = \varepsilon' + i\varepsilon'' \tag{5.1}$$

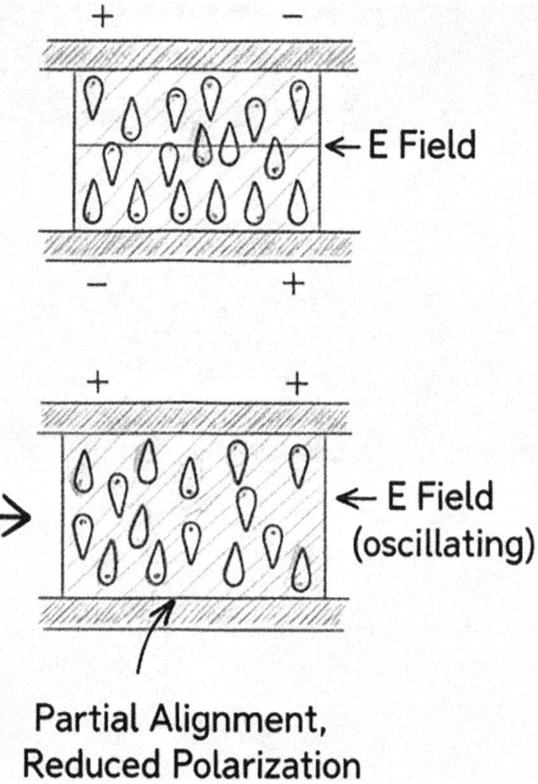

Diagram generated by AI assistance.

Figure 5.1 Illustration of DC field response and AC field response.

The real part (ε') relates to the polarization of the material and its ability to store energy from the electric field. It is associated with the reduction in the speed of the EM wave as it travels through the medium (dispersion). The imaginary part (ε''), also known as the *dielectric loss factor*, relates to the dissipation of energy within the medium, typically as heat. It is a measure of how much the material absorbs the EM wave.

The complex dielectric constant directly determines a material's optical properties through its relationship with the complex refractive index ($\tilde{n}$), also known as the optical constant. For nonmagnetic materials (a common assumption at optical frequencies), the fundamental relationship is

$$\tilde{n}^2 = \tilde{\varepsilon} \tag{5.2}$$

The complex refractive index is also a complex number, defined as follows:

$$\tilde{n} = n + i\kappa \tag{5.3}$$

where n (real part) is *the refractive index*: this determines the phase velocity of light in the material and how much light bends (refraction) at an interface. Similarly, imaginary part is *the extinction coefficient*: this determines the attenuation or absorption of the light's intensity as it propagates through the material.

By equating the real and imaginary parts of Eq. 5.2, we get the explicit relationships that link the electrical response to the optical behavior. The real part of the dielectric constant relates to both the refractive index and the absorption:

$$\varepsilon' = n^2 - \kappa^2 \tag{5.4}$$

The imaginary part of the dielectric constant is directly proportional to both the refractive index and the extinction coefficient, quantifying light absorption:

$$\varepsilon'' = 2n\kappa \tag{5.5}$$

Subsequently, we will discuss dielectric constants in metals and insulators.

5.3 DRUDE ABSORPTION

The Drude model describes how free electrons in a metal or doped semiconductor respond to an EM field. It is the counterpart to the Lorentz model, which treats bound electrons; the Drude model assumes electrons are not bound and can move freely but experience scattering with an average relaxation time. This simple classical picture successfully explains the optical and electrical properties of metals in the infrared and visible ranges.

Momentum-relaxing scattering events (e.g., with phonons or impurities) occur with an average scattering time τ. The damping force is expressed as follows:

$$m\ddot{r} = -eE_{ext} - \frac{m}{\tau}\dot{r} \tag{5.6}$$

$$\dot{r} - \frac{-e}{m\left(\frac{1}{\tau} - i\omega\right)}E \tag{5.7}$$

And the Drude model yields the frequency-dependent conductivity for free electrons:

$$\sigma(\omega) = \frac{ne^2\tau}{m_e(1 - i\omega\tau)} \tag{5.8}$$

The Drude model provides a classical framework for describing electrical conduction in metals, giving rise to the frequency-dependent conductivity. This conductivity is directly connected to the complex relative dielectric function, which describes the material's response to an oscillating electric field, where the imaginary part accounts for energy loss, including that from conduction. The fundamental relationship between the complex relative permittivity and the AC conductivity comes from Maxwell's equations (specifically, the definition of the displacement current) and is given by

$$\tilde{\varepsilon}(\omega)=\varepsilon_1(\omega)+i\varepsilon_2(\omega)=\varepsilon_\infty+\frac{i\sigma(\omega)}{\varepsilon_0\omega} \tag{5.9}$$

$$\tilde{\varepsilon}(\omega)=\varepsilon_\infty-\frac{\omega_{p,\,metal}{}^2}{\omega(\omega-i/\tau)} \tag{5.10}$$

where the plasma frequency for metal is

$$\omega_{p,\,metal}{}^2=\frac{ne^2}{\varepsilon_{0m_e}} \tag{5.11}$$

where ε_∞ is the high-frequency background permittivity (or the permittivity due to bound electrons) and ε_0 is the permittivity of free space, n is the electron density, e is the electron charge, m_e is the electron mass, ω is the angular frequency, and τ is the mean scattering time.

This comprehensive expression demonstrates how the energy dissipation due to the movement of free electrons, as described by the Drude conductivity, manifests as the imaginary part $\varepsilon_2(\omega)$ of the complex permittivity, while the real part $\varepsilon_1(\omega)$ describes the material's ability to store electric field energy. One example is shown in Figure 5.2.

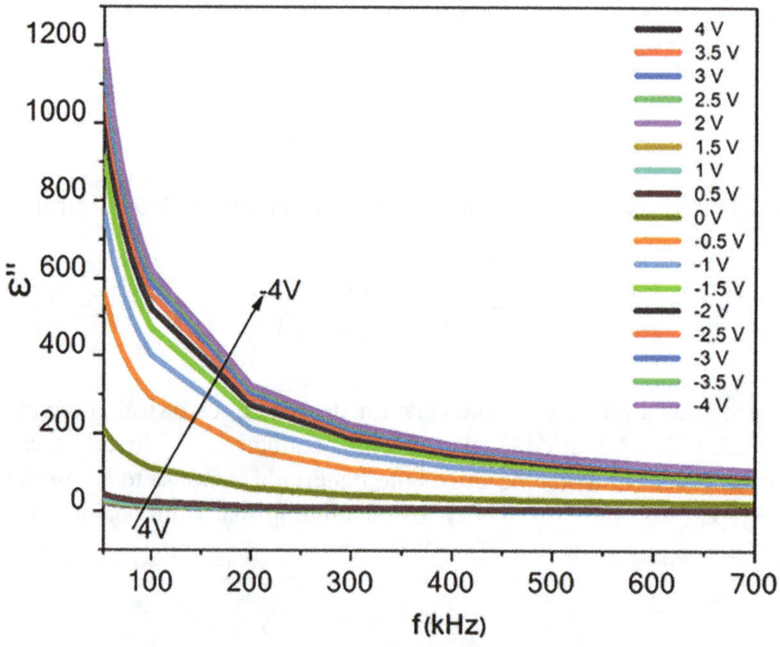

Figure 5.2 Imaginary dielectric constant change with applied bias CC BY 4.0.

5.4 LORENTZ ABSORPTION

The Lorentz model describes how bound electrons in a solid respond to an oscillating EM field and is widely used to understand optical properties such as refractive index, absorption, and dispersion. It treats each electron as a classical damped harmonic oscillator driven by the electric field of light (Figure 5.3) .

In a solid, electrons are not free; they are bound to atoms or ions by restoring forces. When an external electric field $E(t)$ from an incoming light wave interacts with the material, it displaces the bound electrons from equilibrium. The restoring force attempts to pull them back, while damping accounts for energy loss due to scattering or radiation.

This motion determines the polarization of the medium, which in turn influences how light propagates through the solid.

A bound electron of charge $-e$, effective mass m, and natural resonant frequency ω_0 obeys:

$$m\frac{d^2x}{dt^2}+m\gamma\frac{dx}{dt}+m\omega_0^2x=-eE(t) \quad (5.12)$$

where x = displacement of the electron, γ = damping constant, ω_0 = resonant (natural) frequency, and $E(t)$ = applied electric field.

For an optical wave, the electric field is harmonic:

$$E(t)=E_0e^{-i\omega t}$$

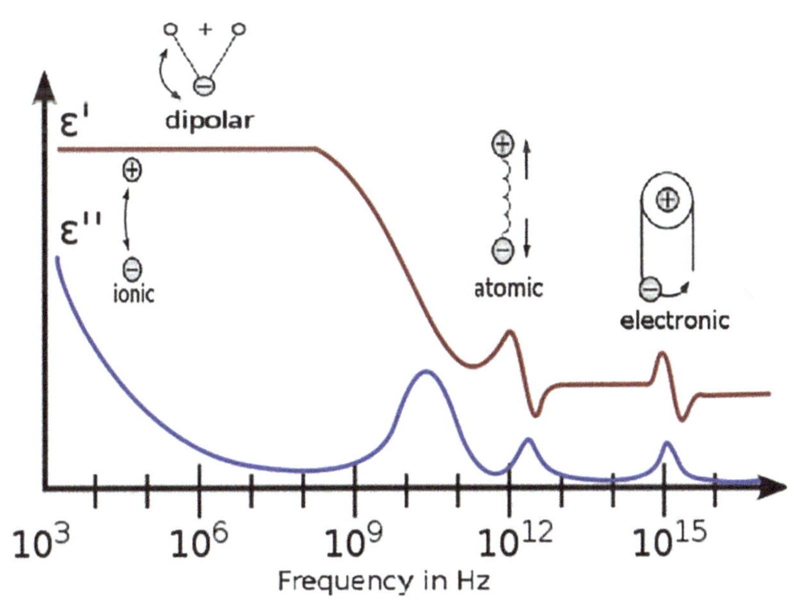

Figure 5.3 Frequency-dependent dielectric constants in the Lorentz model.

Zulkifli Ahmad, Dielectric materials, DOI: 10.5772/50638. OA. Adapted from Application Note 1217-1 (part number 5091-3300E) from Hewlett-Packard Co.

Solving for $x(t)$ gives the steady-state displacement:

$$x(\omega)=\frac{-\frac{e}{m}}{\omega_0^2-\omega^2-i\gamma\omega}E_0e^{-i\omega t} \tag{5.13}$$

The polarization P is the dipole moment per unit volume. For N oscillators per unit volume:

$$P=-Nex(\omega) \tag{5.14}$$

Substituting the expression for $x(\omega)$:

$$P(\omega)=\frac{\frac{Ne^2}{m}}{\omega_0^2-\omega^2-i\gamma\omega}E(\omega) \tag{5.15}$$

The polarization is related to the dielectric function $\varepsilon(\omega)$ by:

$$P=\varepsilon_0\left(\varepsilon(\omega)-1\right)E \tag{5.16}$$

Thus, the *Lorentz dielectric function* becomes

$$\varepsilon(\omega)=1+\frac{Ne^2}{\varepsilon_0 m}\frac{1}{\omega_0^2-\omega^2-i\gamma\omega} \tag{5.17}$$

To separate the real and imaginary parts, multiply the numerator and denominator by the complex conjugate:

$$\varepsilon(\omega)=1+\frac{\frac{Ne^2}{m\left(\omega_0^2-\omega^2+i\gamma\omega\right)}}{(\omega_0^2-\omega^2)^2+(\gamma\omega)^2} \tag{5.18}$$

writing $\varepsilon(\omega)=\varepsilon_1(\omega)+i\varepsilon_2(\omega)$:

Thus,

$$\varepsilon_1(\omega)=1+\frac{\frac{Ne^2}{m\left(\omega_0^2-\omega^2\right)}}{(\omega_0^2-\omega^2)^2+(\gamma\omega)^2} \tag{5.19}$$

$$\varepsilon_2(\omega)=\frac{\frac{Ne^2}{m(\gamma\omega)}}{(\omega_0^2-\omega^2)^2+(\gamma\omega)^2} \tag{5.20}$$

Near the resonance frequency ω_0, dispersion is strong, leading to phenomena such as the following:

The real part $\varepsilon_1(\omega)$ determines the refractive index:

$$n(\omega) = \sqrt{\varepsilon_1(\omega)}$$

For normal dispersion $\omega < \omega_0$:

- $\omega_0^2 - \omega^2 > 0$.
- The numerator is positive.
- Therefore, $\varepsilon_1(\omega)$ increases as ω increases:

$$\frac{dn}{d\omega} > 0 \text{ (normal dispersion)}$$

This is the familiar behavior where the refractive index increases with frequency (or decreases with wavelength).

For anomalous dispersion whereas frequencies just above ω_0:

- $\omega_0^2 - \omega^2 < 0$.
- The numerator becomes negative.
- $\varepsilon_1(\omega)$ decreases rapidly.

$$\frac{dn}{d\omega} < 0 \text{ (anomalous dispersion)}$$

This region is always accompanied by strong absorption ($\omega_2 \neq 0$). Thus, anomalous dispersion occurs only near absorption resonances. For $\omega \gg \omega_0$, both the numerator and denominator scale as ω^2 or ω^4:

$$\varepsilon_1(\omega) \to 1 - \frac{Ne^2}{m\omega^2} \tag{5.21}$$

For the absorption peaks at the resonance frequency, $\omega = \omega_0$.

The imaginary part peaks when the driving frequency matches the resonance:

$$\varepsilon_2(\omega) \propto \frac{\gamma\omega}{(\omega_0^2 - \omega^2)^2 + (\gamma\omega)^2} \tag{5.22}$$

giving strong absorption near ω_0.

At resonance,

$$\varepsilon_2(\omega_0) = \frac{\frac{Ne^2}{m}}{\gamma\omega_0} \tag{5.23}$$

This is the maximum absorption. Physically, the driving frequency matches the natural oscillation frequency of bound electrons, producing large amplitude motion and strong energy dissipation. It is worth noting that absorption is asymmetric around the resonance: for $\omega<\omega_0$, the denominator is dominated by $(\omega_0^2-\omega^2)^2$, so absorption is small; for $\omega>\omega_0$, absorption also drops but more gradually due to the $\gamma\omega$ term. Also, larger damping broadens the absorption peak. This corresponds to stronger scattering or a shorter lifetime of the excited state. In real solids, multiple resonances exist (e.g., phonons, excitons, interband transitions). The dielectric function becomes

$$\varepsilon(\omega) = \varepsilon_\infty + \sum_j \frac{f_j}{\omega_{0j}^2 - \omega^2 - i\gamma_j\omega} \tag{5.24}$$

where f_j = oscillator strength, ω_{0j} = resonance frequencies, γ_j = damping for each mode, and ε_∞ = high-frequency background dielectric constant.

Real solids often include both free and bound carrier responses. The combined model is

$$\varepsilon(\omega) = \varepsilon_\infty - \frac{\omega_p^2}{\omega^2 + \frac{i\omega}{\tau}} + \sum_j \frac{f_j}{\omega_{0j}^2 - \omega^2 - i\gamma_j\omega} \tag{5.25}$$

This is widely used to fit experimental reflectivity or ellipsometry data. The last thing to discuss in the Lorentz model is that we could have a density difference with these bound electrons. In the dilute limit, Lorentz's dielectric function is

$$\varepsilon(\omega) = \varepsilon_c + \frac{\omega_p{}^2}{\omega_0{}^2 - \omega^2 - \frac{i\omega}{\tau}} \tag{5.26}$$

However, the plasma frequency is

$$\omega_p = \sqrt{\frac{4\pi ne^2}{m}}$$

On the other hand, in the dense limit, because of the depolarizing field, the local field correction leads to

$$\varepsilon(\omega) = \varepsilon_c + \frac{\omega_p{}^2}{\omega_0{}^2 - \omega^2 - \frac{i\omega}{\tau} - \frac{1}{3}\omega_p{}^2} \tag{5.27}$$

CHAPTER 6

Transmission and Reflection

Reflection and transmission experiments play a crucial role in understanding the optical properties of materials because they provide direct and nondestructive access to the interaction between electromagnetic waves and matter. By measuring how much of an incident beam is reflected or transmitted across a broad spectral range, one can extract key quantities such as the refractive index, absorption coefficient, dielectric function, and resonance features associated with electronic, vibrational, or excitonic transitions. These measurements are particularly valuable in solids, where complex many-body effects, band structure, and collective modes shape the optical response. Reflection and transmission data form the foundation for modeling using frameworks like the Drude and Lorentz models, enabling precise determination of parameters such as plasma frequency, phonon energies, damping constants, and oscillator strengths. Ultimately, these experiments are indispensable tools in materials research, allowing scientists to probe fundamental excitations, evaluate material quality, guide device design, and verify theoretical predictions.

6.1 THE LAWS OF REFLECTION AND TRANSMISSION

By definition, if light travels from medium ***a*** (refractive index N_a) to medium ***b*** (with thickness r and refractive index N_b) with normal incidence, the reflectance R_{ab} can be defined as follows:

$$|r_{ab}|^2 = R_{ab} = \left|\frac{N_a - N_b}{N_a + N_b}\right|^2 \tag{6.1}$$

$$T_1 = (1 - R_{ab})e^{-\alpha r}(1 - R_{ab}) \tag{6.2}$$

Sum of all the transmittance:

$$\boldsymbol{T} = \frac{(1 - R_{ab})^2 e^{-\alpha r}}{1 - R_{ab}^2 e^{-2\alpha r}} \tag{6.3}$$

Sum of all reflectance:

$$R = R_{ab} + R_{ab}\frac{\left(1-R_{ab}\right)^2 e^{-2\alpha r}}{1-R_{ab}^2 e^{-2\alpha r}} \tag{6.4}$$

For basic spectroscopic reflection and transmission—where the goal is to measure how much light is absorbed, reflected, or transmitted as a function of wavelength—incoherent sources, such as lamps or LEDs, are perfectly sufficient and even preferred because they offer broad spectral coverage. Most conventional optical spectroscopy (UV–Vis–IR) uses incoherent light.

However, coherent light becomes important when the experiment requires phase stability, high spatial or temporal coherence, or precise control over the electromagnetic field. Coherent sources such as lasers are essential for techniques like interferometry, ellipsometry, pump–probe spectroscopy, and nonlinear optical measurements, where detecting phase shifts, interference patterns, or ultrafast dynamics is crucial. Coherence also increases measurement sensitivity, enables very narrow linewidth probing of resonances, and allows accurate characterization of thin films where interference effects dominate (Figure 6.1). Now consider the case of a thin film with refractive index n, thickness d, and complex conductivity is $\sigma_1 + i\sigma_2$. With the following equations, we can invert R and T to find the dielectric constant:

$$T = \frac{4n}{\left(n+1+\frac{4\pi}{c}\sigma_1 d\right)^2 + \left(\frac{4\pi}{c}\sigma_2 d\right)^2} \tag{6.5}$$

$$R = \frac{\left(n-1+\frac{4\pi}{c}\sigma_1 d\right)^2 + \left(\frac{4\pi}{c}\sigma_2 d\right)^2}{\left(2+\frac{4\pi}{c}\sigma_1 d\right)^2 + \left(\frac{4\pi}{c}\sigma_2 d\right)^2} \tag{6.6}$$

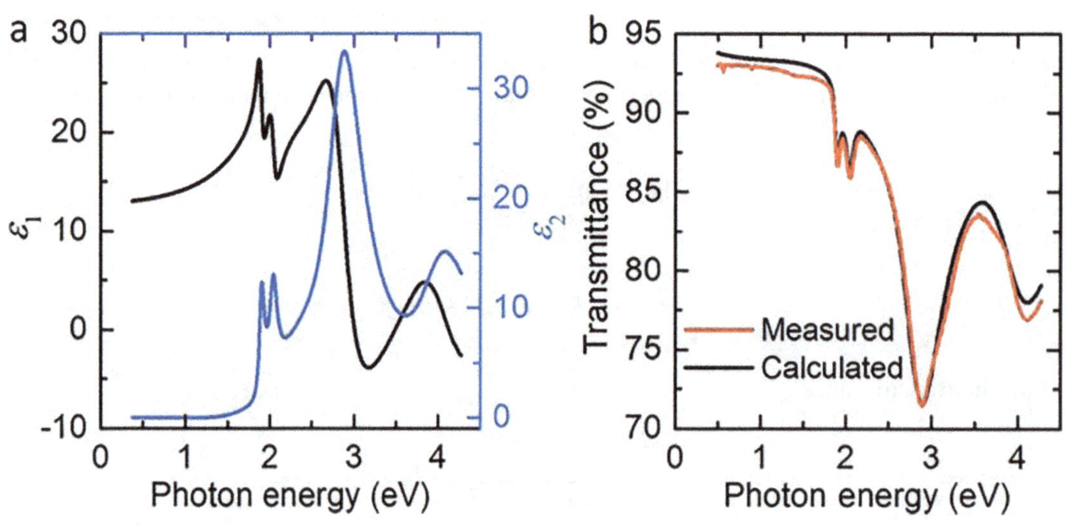

Figure 6.1 Dielectric response and transmission data of MoS2.

Georgy A. Ermolaev et al., npj 2D Mater Appl 4, 21 (2020). OA.

$$T = \frac{4}{\left(2 + \frac{4\pi}{c}\sigma_1 d\right)^2 + \left(\frac{4\pi}{c}\sigma_1 d\right)^2} \tag{6.7}$$

$$R = \frac{2\left(\frac{4\pi}{c}\sigma d\right)^2}{\left(2 + \frac{4\pi}{c}\sigma_1 d\right)^2 + \left(\frac{4\pi}{c}\sigma_1 d\right)^2} \tag{6.8}$$

$$\sigma_1 = \frac{4\pi}{cd}\frac{(1 - R - T)}{T} \tag{6.9}$$

$$\sigma_2 = \frac{4\pi}{cd}\sqrt{\frac{4}{T} - \left(2 + \frac{4\pi}{c}\sigma_1 d\right)^2} \tag{6.10}$$

$$\varepsilon_1(\omega) = \varepsilon_\infty - \frac{\sigma_2(\omega)}{\varepsilon_0 \omega} \tag{6.11}$$

$$\varepsilon_2(\omega) = \frac{\sigma_1(\omega)}{\varepsilon_0 \omega} \tag{6.12}$$

6.2 FRESNEL FORMULA

Although the normal incidence of light can be analyzed with previous chapters, what happens if the incident light wave is not perpendicular to the sample surface? Some of you might recall Snell's law that you have learned before. However, why is Snell's law correct?

Consider light going from point A in medium 1 to point B in medium 2, crossing a flat boundary.

Let

- n_1 = refractive index of medium 1
- n_2 = refractive index of medium 2
- $v_1 = c/n_1$, $v_2 = c/n_2$, speeds of light in the two media
- θ_1 = angle of incidence
- θ_2 = angle of refraction

If the ray hits the boundary at some point x, the total travel time is

$$T = \frac{\text{distance in medium 1}}{v_1} + \frac{\text{distance in medium 2}}{v_2}$$

Write the distances using geometry:

$$T = \frac{\sqrt{x^2 + a^2}}{v_1} + \frac{\sqrt{(d-x)^2 + b^2}}{v_2} \tag{6.13}$$

where a and b are the perpendicular distances from A and B to the interface, respectively.

To find the physical path, take the derivative of T with respect to x and set it to zero:

$$\frac{dT}{dx} = 0$$

Differentiating gives

$$\frac{\sin\theta_1}{v_1} = \frac{\sin\theta_2}{v_2}$$

Substitute $v = c/n$:

$$\frac{n_1 \sin\theta_1}{c} = \frac{n_2 \sin\theta_2}{c}$$

Multiply both sides by c:

$$n_1 \sin\theta_1 = n_2 \sin\theta_2 \tag{6.14}$$

Consider a monochromatic plane wave in a medium:

$$E(\mathbf{r},t) = E_0(\mathbf{r}) e^{i(\varphi(\mathbf{r}) - \omega t)} \tag{6.15}$$

The local spatial phase $\varphi(\mathbf{r})$ increases along the ray. For a homogeneous medium with refractive index n, the (magnitude of the) wave vector is

$$k = \frac{2\pi}{\lambda} = \frac{n\omega}{c}$$

because the wavelength in the medium is $\lambda = \lambda_0/n$ and $k = 2\pi/\lambda$. For a general path C through a possibly inhomogeneous medium where $n = n(\mathbf{r})$ varies slowly on the scale of a wavelength, the local wave number is

$$k(\mathbf{r}) = \frac{n(\mathbf{r})\omega}{c}$$

The total phase accumulated going along a path *C* from point *A* to point *B* is the line integral of the local wave number:

$$\Delta\varphi = \int_C k(\mathbf{r})ds = \frac{\omega}{c}\int_C n(\mathbf{r})ds \tag{6.16}$$

Define the *optical path length L* by scaling phase so that it is proportional to phase:

$$L \equiv \frac{c}{\omega}\Delta\varphi = \int_C n(\mathbf{r})ds \tag{6.17}$$

Thus, $L = \int n\, ds$ measures the phase delay along the path in units of free-space wavelength/phase—it is exactly the geometric length weighted by the local refractive index so that equal OPLs correspond to equal phase shifts.

Further, we can analyze the polarization change after reflection/transmission based on Snell's law and derive the polarization change after reflection:

$$r_s = \frac{n_1 cos\, \theta_i - n_2 cos\, \theta_t}{n_1 cos\, \theta_i + n_2 cos\, \theta_t} \tag{6.18}$$

$$r_p = \frac{n_1 cos\, \theta_t - n_2 cos\, \theta_i}{n_1 cos\, \theta_t + n_2 cos\, \theta_i} \tag{6.19}$$

$$t_s = \frac{2n_1 cos\, \theta_i}{n_1 cos\, \theta_i + n_2 cos\, \theta_t} \tag{6.20}$$

$$t_p = \frac{2n_1 cos\, \theta_i}{n_1 cos\, \theta_t + n_2 cos\, \theta_i} \tag{6.21}$$

The intrinsic impedance of medium *j* is

$$\eta_j = \sqrt{\frac{\mu_j}{\varepsilon_j}} = \frac{\eta_0}{n_j}\left(\mu_j = \mu_0\right) \tag{6.22}$$

For a plane wave in medium *j*,

$$\mathbf{H} = \frac{1}{\eta_j}\hat{k} \times \mathbf{E} \tag{6.23}$$

So the tangential component of **H** at the interface will pick up a factor cos θ (geometry) and a factor $1/\eta_j$ (material). If we choose coordinates, plane of incidence = xz-plane; interface $z = 0$. For **s** polarization, the electric field is along y (perpendicular to the plane):

Incident field (amplitude E_{0i}):

$$\mathbf{E}_i = E_{0i}\hat{y}e^{i(\mathbf{k}_i\cdot\mathbf{r}-\omega t)},\ \mathbf{k}_i = k_1\left(\sin\theta_i, 0, \cos\theta_i\right) \tag{6.24}$$

Reflected:

$$\mathbf{E}_r = E_{0r}\,\hat{y}\,e^{i(\mathbf{k}_r\cdot\mathbf{r}-\omega t)},\ \mathbf{k}_r = k_1\left(\sin\theta_i, 0, -\cos\theta_i\right) \tag{6.25}$$

Transmitted:

$$\mathrm{E}_t = E_{0t}\hat{y}e^{i(\mathbf{k}_t\cdot\mathbf{r}-\omega t)},\ \mathrm{k}_t = k_2\left(\sin\theta_t, 0, \cos\theta_t\right) \tag{6.26}$$

With boundary condition—tangential E at $z = 0$,

$$E_{0i} + E_{0r} = E_{0t}$$

Now compute the tangential component of **H** (the component along x at the boundary). Use $(1/\eta)\,\hat{k}\times\mathbf{E}$. For the incident wave,

$$\hat{k}_i \times \hat{y} = -\cos\theta_i\,\hat{x} + \sin\theta_i\,\hat{z}$$

so the tangential (x) component magnitude is

$$H_{ix} = -\frac{\cos\theta_i}{\eta_1}E_{0i} \tag{6.27}$$

(Sign conventions will produce the standard algebraic signs later.)

Similarly, for the reflected wave, the x-component of $\mathbf{H}_r$ is $+\dfrac{\cos\theta_i}{\eta_1}E_{0r}$ (opposite sign due to reflected propagation direction), and for transmitted,

$$H_{tx} = -\frac{\cos\theta_t}{\eta_2}E_{0t} \tag{6.28}$$

Boundary condition—tangential H continuous at $z = 0$:

$$H_{ix} + H_{rx} = H_{tx}.$$

Put the signs/magnitudes in

$$-\frac{\cos\theta_i}{\eta_1}E_{0i}+\frac{\cos\theta_i}{\eta_1}E_{0r}=-\frac{\cos\theta_t}{\eta_2}E_{0t} \tag{6.29}$$

Multiply by –1 and rearrange

$$\frac{\cos\theta_i}{\eta_1}\left(E_{0i}-E_{0r}\right)=\frac{\cos\theta_t}{\eta_2}E_{0t} \tag{6.30}$$

Now, for the reflection amplitude, $r_s \equiv E_{0r}/E_{0i}$, and for transmission amplitude, $t_s \equiv E_{0t}/E_{0i}$.

From Eq. 6.29: $E_{0t}=E_{0i}+E_{0r}=E_{0i}\left(1+r_s\right)$. Substitute into Eq. 6.30:

$$\frac{\cos\theta_i}{\eta_1}\left(1-r_s\right)=\frac{\cos\theta_t}{\eta_2}\left(1+r_s\right) \tag{6.31}$$

Solve for r_s:

$$\cos\theta_i/\eta_1-\left(\cos\theta_i/\eta_1\right)r_s=\left(\cos\theta_t/\eta_2\right)+\left(\cos\theta_t/\eta_2\right)r_s$$

$$r_s\left[-\frac{\cos\theta_i}{\eta_1}-\frac{\cos\theta_t}{\eta_2}\right]=\frac{\cos\theta_t}{\eta_2}-\frac{\cos\theta_i}{\eta_1}$$

$$r_s=\frac{\dfrac{\cos\theta_t}{\eta_2}-\dfrac{\cos\theta_i}{\eta_1}}{\dfrac{\cos\theta_t}{\eta_2}+\dfrac{\cos\theta_i}{\eta_1}} \tag{6.32}$$

Replace $1/\eta_j=\eta_0/\eta_j$ (since $\eta_j=\eta_0/\eta_j$) and multiply the numerator and denominator by η_0 to get the familiar form:

$$r_s=\frac{n_1\cos\theta_i-n_2\cos\theta_t}{n_1\cos\theta_i+n_2\cos\theta_t} \tag{6.33}$$

and

$$t_s=\frac{2n_1\cos\theta_i}{n_1\cos\theta_i+n_2\cos\theta_t} \tag{6.34}$$

(using $t_s=1+r_s$ times the ratio of fields or from direct algebra). These Fresnel's equations give the *complex numbers* t_s and r_s for *s* polarization transmission and reflection change after incidence (Figure 6.2).

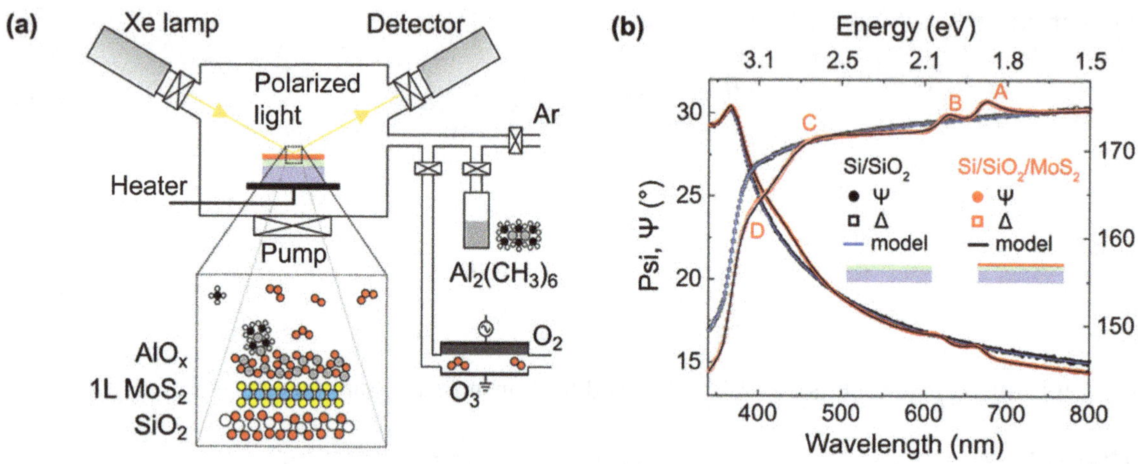

Figure 6.2 Ellipsometry measurements for in situ MoS2 growth.

Adv Mater Interfaces 2023, 10, 2202429.Wiley. OA.

6.3 ELLIPSOMETRY

The fundamental ellipsometry equation is

$$\rho = \frac{r_p}{r_s} = \tan \Psi e^{i\Delta} \tag{6.35}$$

where Ψ is the amplitude ratio (strength difference between p and s) and Δ is the phase difference between p and s after reflection.

Thus,

$$\tan \Psi = \left| \frac{r_p}{r_s} \right| \tag{6.36}$$

$$\Delta = \arg(r_p) - \arg(r_s) \tag{6.37}$$

As demonstrated, Fresnel's equations give the *complex numbers* r_p and r_s. Ellipsometry measures how p and s change relative to each other during reflection. So ellipsometry essentially *inverts Fresnel's equations* to find:

6.4 ELLIPSOMETRY ON THIN FILMS (MULTIPLE REFLECTION CASE)

For a thin film of thickness d on a substrate, Fresnel equations combine using multiple beam interference. The effective reflection coefficient becomes the following:

For *s*-polarization:

$$r_s = \frac{r_{01}^s + r_{12}^s e^{2i\beta}}{1 + r_{01}^s r_{12}^s e^{2i\beta}} \tag{6.38}$$

For p-polarization:

$$r_p = \frac{r_{01}^p + r_{12}^p e^{2i\beta}}{1 + r_{01}^p r_{12}^p e^{2i\beta}} \tag{6.39}$$

where r_{01} is the Fresnel coefficient at air → film, r_{12} is the Fresnel coefficient at film → substrate, and $\beta = \frac{2\pi d}{\lambda} n_f \cos\theta_f$ is the *film phase thickness.*

This introduces both *thickness* and *refractive index* into the measurable signal.

CHAPTER 7

Absorption of Light in Solids

Absorption of light in solids arises from the ability of electrons and, in some cases, lattice vibrations to take up energy from an electromagnetic wave. The microscopic origin of absorption depends strongly on the electronic structure of the material. Metals, semiconductors, and insulators each exhibit distinct absorption behaviors because their band structures and carrier densities differ dramatically.

7.1 THE ABSORPTION COEFFICIENT

Here, we summarize all the absorption coefficients together, and hopefully, after learning this chapter, all these coefficients will make sense to us.

$$\alpha_{ab}(\omega) = \frac{2\omega k_i(\omega)}{c} \tag{7.1}$$

Metal:

$$\alpha_{ab}(\omega) \sim \sqrt{\omega} \tag{7.2}$$

Semiconductor:

$$\alpha_{ab}(\omega) \sim \omega^{-2} \tag{7.3}$$

Direct interband:

$$\alpha_{ab}(\omega) \sim \frac{\sqrt{(\hbar\omega - E_g)}}{\hbar\omega} \tag{7.4}$$

Forbidden transition:

$$\alpha_{ab}(\omega) \sim \frac{(\hbar\omega - E_g)^{3/2}}{\hbar\omega} \tag{7.5}$$

Indirect interband:

$$\alpha_{ab}(\omega) \sim (\hbar\omega - E_g \pm \hbar\omega_q)^2 \tag{7.6}$$

7.2 FREE CARRIER ABSORPTION IN METALS

Metals contain a very high density of free electrons. These electrons can respond to low-frequency electromagnetic waves almost as a plasma, leading to characteristic metallic behavior:

- **Free-carrier absorption (Drude absorption):** Conduction electrons can absorb photons by being accelerated and scattering with phonons or impurities. The Drude model describes this with a dielectric function:

$$\varepsilon(\omega) = \varepsilon_\infty - \frac{\omega_p^2}{\omega(\omega + i\gamma)} \tag{7.7}$$

where ω_p is the plasma frequency.

- At *frequencies below the plasma frequency* ($\omega < \omega_p$), metals exhibit strong reflection and absorption because the real part of ε is negative, and electromagnetic waves are strongly damped within a very short *skin depth* (tens of nanometers).

Overall, metals absorb strongly at low frequencies due to free electrons and at higher frequencies due to interband transitions.

$$k(\omega) = \sqrt{\frac{2\pi n e^2 \tau}{m^* \omega}} \tag{7.8}$$

7.3 FREE CARRIER ABSORPTION IN SEMICONDUCTORS

Insulators have a large bandgap, often several electron volts. The valence band is fully occupied, and the conduction band is empty. Because a photon must have energy $E = hv \geq E_g$

to excite an electron across the gap, insulators show no absorption for photon energies below the bandgap (transparent in visible or infrared depending on the size of the gap); sharp absorption onset at $E = E_g$, often producing a steep rise in the absorption coefficient; and phonon absorption in the infrared (IR): lattice vibrations can absorb low-energy photons (in the mid-IR), producing strong absorption bands described by Lorentz oscillators.

On the other hand, semiconductors occupy the middle ground between metals and insulators. They have a moderate *bandgap* (0.1–3 eV), a moderate density of free carriers (depending on doping), and strong excitonic and interband transitions.

The following three main absorption mechanisms are important:

a) Interband absorption

When photon energy exceeds the bandgap E_g, electrons are excited from the valence to the conduction band.

- In direct-gap semiconductors (GaAs, InP), absorption rises sharply:

$$\alpha(\hbar\omega) \propto \sqrt{\hbar\omega - E_g} \tag{7.9}$$

- In indirect-gap semiconductors (Si, Ge), phonon assistance is required, giving a slower absorption onset. Interband indirect absorption describes the optical transition of an electron from the valence band to the conduction band in materials, where the energy minimum of the conduction band and the maximum of the valence band occur at different crystal momentum ($\vec{k}$) values. This process is characteristic of indirect bandgap semiconductors like silicon. Since the incoming photon provides significant energy ($\hbar\omega$) but negligible momentum, the electron must interact with the crystal lattice via a phonon (which has energy $E_p = \hbar\omega_p$ and momentum k_p to satisfy both energy and momentum conservation laws simultaneously). The governing conservation equations are as follows:

- Energy: $\hbar\omega = E_g \pm \hbar\omega_p$
- Momentum: $\vec{k}_{final} - \vec{k}_{initial} = \vec{k}_p$

The ± accounts for the absorption (+) or emission (−) of a phonon, making this a second-order process with a characteristically weak, temperature-dependent absorption coefficient near the band edge.

b) Excitonic absorption

Near the absorption edge, electron–hole pairs can form bound states called excitons.

These produce sharp peaks slightly below the bandgap energy.

c) Free-carrier (Drude) absorption

Doped semiconductors contain free electrons or holes:

- These carriers absorb low-energy photons, especially in the THz–IR range.
- Described by the Drude form of $\varepsilon(\omega)$.

Semiconductors therefore combine features of both insulators (bandgap absorption) and metals (free-carrier absorption when doped).

$$\varepsilon_2(\omega) = 2n(\omega)k(\omega) \approx 2\sqrt{\varepsilon_0}\,k(\omega) = \frac{\varepsilon_0\omega_p^2\tau^2}{\omega^3\tau^3} \tag{7.10}$$

d) Direct interband

Fermi's golden rule leads to

$$\alpha_{ab}(\omega) \sim \frac{\sqrt{(\hbar\omega - E_g)}}{\hbar\omega} \tag{7.11}$$

III

Modern Interpretation for Optical Properties of Solids

CHAPTER 8

Kramers–Kronig Relations and Sum Rules

The *Kramers–Kronig (KK) relations* are fundamental mathematical theorems in solid-state optics that establish a powerful, inescapable link between the real and imaginary parts of any response function, such as the dielectric function ($\in$) or the complex refractive index (n). These relations arise from the principle of causality—the physical effect cannot precede its cause. In the context of light interacting with a solid, this means that the material's dispersion (how the speed of light changes with frequency, characterized by the real part of the index, ε_1) must be entirely determined by its absorption (how much light is attenuated, characterized by the imaginary part, ε_2, or the extinction coefficient), and vice versa. The KK relations are represented by principal value integrals (what do these integrals remind you of mathematically?):

$$\varepsilon_1(\omega)=1+\frac{2}{\pi}P\int_0^{\infty}\frac{\omega' k(\omega')}{\omega'^2-\omega^2}d\omega' \tag{8.1}$$

$$\varepsilon_2(\omega)=-\frac{2\omega}{\pi}P\int_0^{\infty}\frac{n(\omega')-1}{\omega'^2-\omega^2}d\omega' \tag{8.2}$$

The practical importance of the KK relations in experimental physics and materials science is immense as they allow scientists to determine a material's complete optical profile from a single measurable spectrum. It is often far easier to measure the absorption spectrum (e.g., using transmission or reflection spectroscopy) across a wide range of frequencies than it is to measure both the absorption and dispersion simultaneously and accurately. By measuring the imaginary part of the optical constants (absorption) over all frequencies and applying the KK transformation, researchers can theoretically calculate the real part (dispersion) without further experimentation. This eliminates redundant measurements and provides a vital self-consistency check for experimental data, ensuring that the measured optical properties obey fundamental physical laws.

8.1 KRAMERS–KRONIG INTEGRALS

Other than the most common forms for dielectric constants,

$$\varepsilon_1(\omega)-1=\frac{2}{\pi}\boldsymbol{P}\int_0^{\infty}\frac{\omega'\varepsilon_2(\omega')}{\omega'^2-\omega^2}d\omega' \tag{8.3}$$

$$\varepsilon_2(\omega)=-\frac{2}{\pi}\boldsymbol{P}\int_0^{\infty}\frac{\omega'\varepsilon_1(\omega')}{\omega'^2-\omega^2}d\omega' \tag{8.4}$$

Two more forms should be introduced: the *optical conductivity form* and the *refractive index form*.

For the complex conductivity,

$$\tilde{\sigma}(\omega)=\sigma_1(\omega)+i\sigma_2(\omega)$$

$$\sigma_2(\omega)=-\frac{2\omega}{\pi}P\int_0^{\infty}\frac{\sigma_1(\omega')}{\omega'^2-\omega^2}d\omega' \tag{8.5}$$

$$\sigma_1(\omega)=\frac{2}{\pi}P\int_0^{\infty}\frac{\omega'\sigma_2(\omega')}{\omega'^2-\omega^2}d\omega' \tag{8.6}$$

For the complex refractive index,

$$\tilde{n}(\omega)=n(\omega)+i\kappa(\omega)$$

$$n(\omega)-1=\frac{2}{\pi}P\int_0^{\infty}\frac{\omega'\kappa(\omega')}{\omega'^2-\omega^2}d\omega' \tag{8.7}$$

$$\kappa(\omega)=-\frac{2\omega}{\pi}P\int_0^{\infty}\frac{n(\omega')-1}{\omega'^2-\omega^2}d\omega' \tag{8.8}$$

8.2 SUM RULE

The *optical sum rule* is a fundamental principle in the study of the optical properties of materials. It expresses the idea that the total strength of all possible optical transitions in a material is conserved. In other words, the *integrated optical response* over all frequencies depends only on fundamental quantities such as the electron density and mass, and not on the detailed interactions or structure of the material. Mathematically, for the real part of the optical conductivity $\sigma_1(\omega)$, the sum rule can be written as follows:

$$\int_0^{\infty}\sigma_1(\omega)d\omega=\frac{\pi ne^2}{2m} \tag{8.9}$$

where n is the electron density, e is the electron charge, and m is the electron mass (or effective mass in solids). This implies that while interactions, temperature changes, or phase transitions can redistribute spectral weight across different frequencies, the total "area" under the optical conductivity curve remains constant. This makes the sum rule a powerful tool for understanding changes in electronic structure and identifying features such as free-carrier contributions, interband transitions, or effects of strong correlations.

A classical derivation of the sum rule can be made using the Drude–Lorentz model of electrons bound in a harmonic potential. Consider an electron of charge e and mass m subject to an oscillating electric field $E(t) = E_0 e^{-i\omega t}$. The equation of motion is

$$m\frac{d^2x}{dt^2} + m\gamma\frac{dx}{dt} + m\omega_0^2 x = -eE_0 e^{-i\omega t} \tag{8.10}$$

where ω_0 is the natural frequency of the oscillator and γ is the damping rate. Solving for the velocity $v = dx/dt$, the complex conductivity is

$$\sigma(\omega) = \frac{ne^2}{m}\frac{i\omega}{\omega_0^2 - \omega^2 + i\gamma\omega} \tag{8.11}$$

The real part $\sigma_1(\omega)$ describes absorption. Integrating $\sigma_1(\omega)$ over all frequencies gives

$$\int_0^\infty \sigma_1(\omega)\,d\omega = \frac{\pi ne^2}{2m} \tag{8.12}$$

which reproduces the sum rule. This derivation shows that the sum rule is independent of the specific resonance frequency ω_0 or damping γ; it only depends on fundamental electron parameters.

In practice, the sum rule is widely used to analyze experimental optical spectra. For metals, most of the spectral weight appears at low frequencies (the Drude peak), while for insulators and semiconductors, it shifts to higher-frequency interband transitions. In strongly correlated systems, such as high-temperature superconductors, spectral weight can redistribute dramatically across different energy scales, but the sum rule ensures that the total integrated weight remains fixed. This makes it a powerful diagnostic for understanding electron dynamics, effective mass, and phase transitions in materials, as depicted in Figure 8.1.

How about discrete resonant frequencies? The sum rule is a principle that imposes a constraint on the total "strength" of all possible transitions or resonances in a system. For discrete resonant frequencies, it usually refers to discrete energy levels (like in atoms or molecules) and their corresponding oscillator strengths. *Oscillator strength* f_n is a dimensionless number that quantifies the strength of the transition from a ground state to the n-th excited state. *Discrete resonant frequencies* ω_n correspond to energy differences between the ground state and the n-th excited state:

$$\hbar\omega_n = E_n - E_0$$

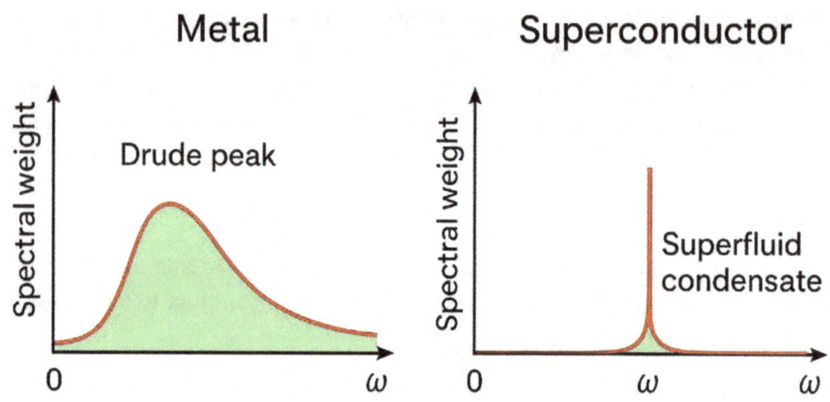

Figure 8.1 Metal and superconductor showing distinct spectral weight differences.

Diagram generated by AI assistance.

The *Thomas–Reiche–Kuhn (TRK) sum rule* is a classic example:

$$\sum_n f_n = N \tag{8.13}$$

where N is the number of electrons (for an atomic system). This states that the sum of oscillator strengths over all discrete transitions is constant, independent of the details of the system. If we describe the system's susceptibility $\chi(\omega)$ using a sum over Lorentzian-like resonances,

$$\chi(\omega) = \sum_n \frac{f_n}{\omega_n^2 - \omega^2 - i\gamma_n \omega} \tag{8.14}$$

where γ_n is the damping.

Physically, this means that even if individual resonances shift or broaden, the total "strength" of all possible transitions is fixed. This is important in spectroscopy because it ensures energy conservation and consistency with quantum mechanics. Let us review one example: simple 1D quantum harmonic oscillator (Figure 8.2).

- Discrete energy levels: $E_n = \hbar\omega_0\left(n + \frac{1}{2}\right)$.
- Transition frequencies: $\omega_{n,\,n+1} = \omega_0$.
- Oscillator strengths f_n can be calculated using matrix elements of the position operator x (dipole approximation):

$$f_n = \frac{2m\omega_{n0}}{\hbar} \left|\langle n|x|0\rangle\right|^2 \tag{8.15}$$

- Sum over all $_n$ satisfies the sum rule: $\sum_n f_n = 1$ for one electron.

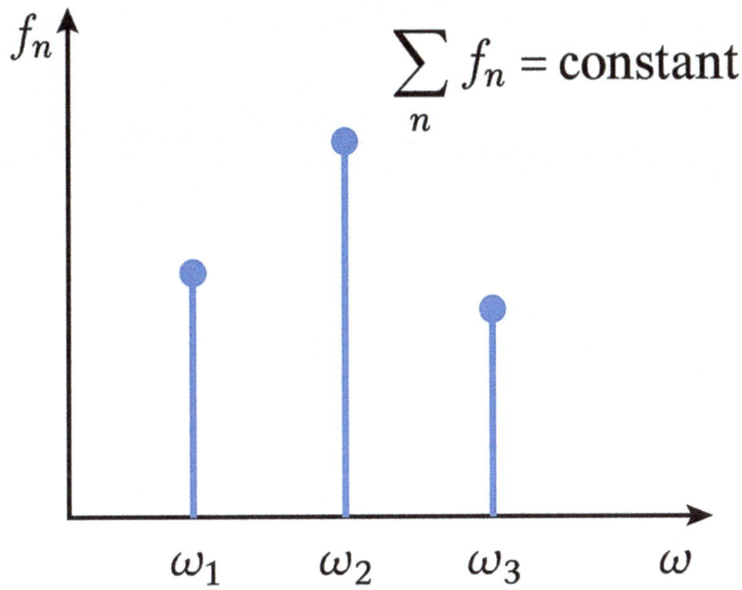

Figure 8.2 1D simple harmonic sum rule representation.

Graph generated with the assistance of AI.

CHAPTER 9

Quasiparticles with Light–Matter Interaction

9.1 PHONONS

A phonon is a quantized collective excitation of the atoms in a crystalline solid, representing a normal mode of lattice vibration in which atoms oscillate coherently about their equilibrium positions. Instead of describing each atom's motion individually, lattice vibrations are treated as waves characterized by a frequency and wave vector, and when these vibrational modes are quantized, their energy is carried in discrete packets called phonons. Phonons play a central role in determining a material's thermal, mechanical, and optical properties, governing phenomena such as heat conduction, sound propagation, thermal expansion, and the interaction of light and electrons with the lattice. Next, we will describe a classical description of how phonons couple to light via polarization, and they form the physical basis of Raman scattering and infrared activity. I will explain them step by step and explicitly connect each term to phonons. The macroscopic polarization P induced by an electric field E is $P = \alpha E$, where α is the *polarizability* and P is the *dipole moment per unit volume.*

Assume an incident monochromatic light wave:

$$E(t) = E_0 \sin(\omega_0 t) \tag{9.1}$$

where ω_0 is the optical frequency of the incoming photon.

A phonon is a collective lattice vibration with frequency ω_q. The ionic displacement associated with the phonon *changes the electronic cloud*, so the polarizability is not constant but *oscillates in time*:

$$\alpha(t) = \alpha_0 + \alpha_1 \sin(\omega_q t) \tag{9.2}$$

where α_0 is the equilibrium polarizability, α_1 is the change in polarizability due to phonon displacement, and ω_q is the phonon frequency.

$$P(t) = E_0\alpha_0 \sin(\omega_0 t) + E_0\alpha_1 \sin(\omega_q t)\sin(\omega_0 t) \tag{9.3}$$

$$P = E_0\left[\alpha_0 \sin \omega_0 t + \frac{1}{2}\alpha_1 \cos\left(\omega_0 - \omega_q\right)t - \frac{1}{2}\alpha_1 \cos\left(\omega_0 + \omega_q\right)t\right] \tag{9.4}$$

Quantum mechanically, Raman scattering is an *inelastic light scattering* process that provides information on a material's vibrational modes (phonons in solids). The scattered light can have lower energy (Stokes) or higher energy (anti-Stokes) than the incident light, with selection rules linking the process to high-symmetry points in the material's *k*-space (Brillouin zone).

Most light interacting with a material is elastically scattered (Rayleigh scattering) with no change in frequency. However, a small fraction undergoes Raman scattering, where energy is exchanged between the incoming photon and the material's molecular vibrations or phonons. This results in scattered light at a different frequency (wavelength) from the incident light. Figure 9.1 illustrates the response of the Raman experiment, and we can focus on these two branches:

Stokes Scattering: In this process, the incident photon transfers a portion of its energy to the molecule, exciting it from a lower vibrational energy state (usually the ground state) to a higher one. The scattered photon thus has a lower energy (and longer wavelength) than the incident photon. Stokes scattering is the most common and intense type of Raman scattering at normal temperatures because most molecules are in their ground vibrational state according to the Boltzmann distribution.

Anti-Stokes Scattering: This occurs when the incident photon interacts with a molecule that is already in an excited vibrational state. The photon gains energy from the molecule, which relaxes to a lower-energy vibrational state (usually the ground state). The scattered photon consequently has higher energy (and shorter wavelength) than the incident photon. Anti-Stokes lines are generally much weaker than Stokes lines because the population of excited vibrational states is lower at typical temperatures.

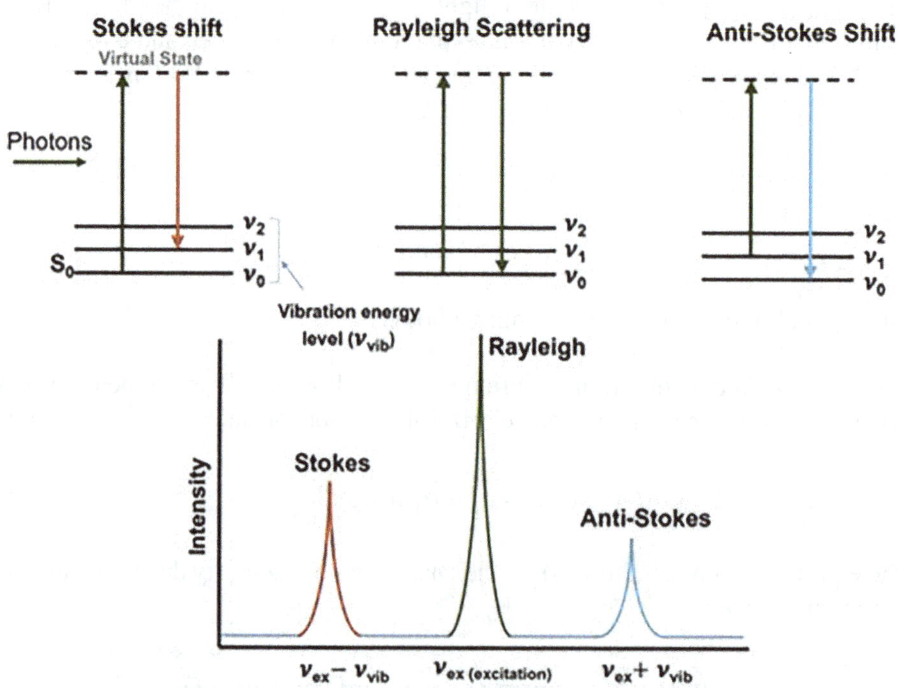

Figure 9.1 Raman scattering with Stokes shift and anti-Stokes shift showing energy level dependence. *Sci Rep 10*, 11692 (2020). OA.

In crystalline solids, the analysis of Raman scattering involves considering the crystal's symmetry and the behavior of phonons within its reciprocal space, known as k-space or the Brillouin zone. The wave vector of the phonon (q) must be such that the total momentum (wave vector) is conserved during the scattering event:

$$k_{incident} = k_{scattered} + q$$

For typical first-order Raman scattering experiments using visible or near-infrared lasers, the wave vectors of the incident and scattered photons are very small compared to the dimensions of the Brillouin zone. Consequently, the phonons involved in the first-order Raman process must have a wave vector extremely close to the center of the Brillouin zone, which is referred to as the *Γ point.*

Second-Order Raman Scattering: The most common way to access phonons away from the Γ point is through second-order Raman scattering (also known as two-phonon processes). In this scenario, two phonons are created (Stokes) or destroyed (anti-Stokes), or one is created and one is destroyed.

Momentum Conservation: The crucial difference is that the wave vectors of the two phonons must sum to the photon's wave vector. This condition allows the involvement of phonons with large but opposite wave vectors from anywhere in the Brillouin zone, including high-symmetry points like the K and M points. The resulting spectrum in second-order Raman scattering often provides information that resembles the phonon density of states (PDOS) integrated over the entire Brillouin zone, rather than sharp peaks corresponding to single modes.

Lastly, we can compare Raman active phonons vs infrared-active phonons:

- **Raman-active phonons:**

$$\frac{d\alpha}{dQ} \neq 0$$

(Polarizability changes with phonon coordinate.)

- **Infrared-active phonons:**

$$\frac{dP}{dQ} \neq 0$$

(Dipole moment changes directly.)

9.2 EXCITON

An exciton is a neutral, mobile "quasi-particle" in materials like semiconductors, consisting of an electron and a positively charged "hole" (an electron vacancy) bound together by their mutual electrostatic attraction (Coulomb force), formed when light excites an electron to a higher energy level, leaving a hole behind. A *Frenkel exciton* represents a localized excitation on lattice/site n by $|n\rangle$. A minimal Frenkel Hamiltonian is

$$\widehat{\boldsymbol{H}} = \sum_{n} \boldsymbol{E}_0 |\boldsymbol{n}\rangle\langle\boldsymbol{n}| + \sum_{n\neq m} J_{nm} |\boldsymbol{n}\rangle\langle\boldsymbol{m}| \quad (9.5)$$

where E_0 is the on-site excitation energy and J_{nm} is the excitation (resonant) transfer integral between sites *n, m*. A Fourier transform to Bloch states $|\mathrm{k}\rangle = \frac{1}{\sqrt{N}}\sum_n e^{i\mathrm{k}\cdot\mathrm{R}_n}|n\rangle$ gives the dispersion:

$$E(\mathbf{k}) = E_0 + \sum_R J(R)e^{i\mathrm{k}\cdot\mathrm{R}} \tag{9.6}$$

showing a band of exciton states; the localization length is of the order of the lattice constant when *J* is small and Coulomb binding is strong. Now, let us look at another exciton category, *Wannier–Mott (hydrogen-like) excitons*. We start with the relative-motion Schrödinger picture:

$$\left[-\frac{\hbar^2}{2\mu}\nabla_{\mathbf{r}}^2 - \frac{e^2}{4\pi\varepsilon_0\varepsilon_r r}\right]\psi(\mathbf{r}) = E\psi(\mathbf{r}) \tag{9.7}$$

where $\mathbf{r} = \mathbf{r}_e - \mathbf{r}_h$ is the relative coordinate, $\mu = (m_e^* m_h^*)/(m_e^* + m_h^*)$ is the reduced effective mass, and ε_r is the relative dielectric constant. The hydrogenic eigenenergies (3D) are

$$E_n = -\frac{R^*}{n^2},\ R^* = \frac{\mu e^4}{2(4\pi\varepsilon_0\varepsilon_r)^2\hbar^2} \tag{9.8}$$

and the effective Bohr radius is

$$a^* = \frac{4\pi\varepsilon_0\varepsilon_r\hbar^2}{\mu e^2} \tag{9.9}$$

Frenkel and Wannier–Mott excitons differ mainly in how tightly the electron–hole pair is bound and how far it extends in a material. Frenkel excitons are strongly bound, highly localized excitations that typically occur in molecular crystals, organic semiconductors, and low-dielectric materials; the electron and hole remain close—often within a single molecule or unit cell—because screening is weak, giving these excitons small radii and large binding energies. In contrast, Wannier–Mott excitons form in conventional inorganic semiconductors with high dielectric constants, where strong screening reduces the Coulomb attraction and allows the electron–hole pair to spread over many lattice constants. As a result, Wannier–Mott excitons are much larger, more delocalized, and have much smaller binding energies, often described by a hydrogen-like effective-mass model.

The *exciton emission rate* refers to the probability per unit time that an exciton recombines and emits a photon. It is the inverse of the exciton's radiative lifetime and is typically defined within the framework of spontaneous emission.

For an exciton state $|X\rangle$, the radiative emission rate (spontaneous emission rate) is

$$\Gamma_{\mathrm{rad}} = \frac{1}{\tau_{\mathrm{rad}}} \tag{9.10}$$

where τ_{rad} is its radiative lifetime.

More formally, using Fermi's golden rule, the emission rate is

$$\Gamma_{\text{rad}} = \frac{2\pi}{\hbar}\sum_{\mathbf{k},\lambda}\left|\left\langle G;1_{\mathbf{k},\lambda}\left|\widehat{H}_{\text{int}}\right|X\right\rangle\right|^2 \delta\left(E_X - \hbar\omega_{\mathbf{k}}\right) \tag{9.11}$$

where $|X\rangle$ is the exciton state, $\left|G;1_{\mathbf{k},\lambda}\right\rangle$ is the ground state plus one emitted photon, $\widehat{H}_{\text{int}}$ is the light–matter interaction Hamiltonian, λ is the photon polarization, and δ enforces energy conservation.

9.3 MAGNON

Magnons are quantized spin wave excitations whose energies are obtained by mapping the Heisenberg Hamiltonian to bosonic operators (via Holstein–Primakoff or Dyson–Maleev), yielding a dispersion $\omega(\mathbf{K})$ that is quadratic for ferromagnets and linear for antiferromagnets.

Magnon frequencies measured by Brillouin light scattering (BLS) originate from the spin wave dispersion of magnetic materials. For a ferromagnet in the Heisenberg model, the magnon frequency is

$$\hbar\omega(\mathbf{k}) = 2JSz\left(1-\gamma_{\mathbf{k}}\right) \tag{9.12}$$

which reduces in the long-wavelength limit to

$$\hbar\omega \approx Dk^2 \tag{9.13}$$

For a two-sublattice antiferromagnet, the dispersion becomes

$$\hbar\omega(\mathbf{k}) = 2JS\sqrt{1-\gamma_{\mathbf{k}}^2} \tag{9.14}$$

BLS spectroscopy is an *inelastic light-scattering technique* used to study *acoustic phonons* (sound waves) and *magnons* (spin waves) in materials.

When monochromatic light interacts with thermally excited excitations in a medium, it scatters with a slight frequency shift (typically GHz order), which carries information about the mechanical, optical, or magnetic properties of the material.

This can be written as follows:

$$\omega_s = \omega_i \pm \Omega$$

where ω_i is the incident light frequency, ω_s is the scattered light frequency, and Ω is the frequency of the magnon.

For a light of wave vector $\mathbf{k}$ interacting with a phonon of wave vector $\mathbf{q}$:

$$\mathbf{k}_s = \mathbf{k}_i \pm \mathbf{q}$$

In backscattering geometry (most common),

$$|\mathbf{k}_s| = |\mathbf{k}_i| = \frac{2\pi n}{\lambda}$$

where n is the refractive index of the material and λ is the vacuum wavelength of the incident light.

Brillouin shift in frequency (Hz):

$$\nu_B = \frac{\Omega}{2\pi} = \frac{2nv}{\lambda}$$

This is the *central equation of BLS spectroscopy.*

9.3.1 Brillouin Light Scattering for Magnons (Magnetic Materials)

When scattering from magnons (spin waves), the frequency shift corresponds to the magnon frequency:

$$\Omega = \omega_{\text{magnon}}(q)$$

Spin wave dispersion (Damon–Eshbach mode, thin film):

$$\omega(q) = \gamma\sqrt{(H + Dq^2)(H + Dq^2 + M_s)} \tag{9.15}$$

where γ is the gyromagnetic ratio, H is the applied field, D is the exchange stiffness, and M_s is the saturation magnetization.

Here, BLS measures the *spin wave frequency* at the wave vector

$$q = \frac{4\pi n}{\lambda}\sin(\theta/2) \tag{9.16}$$

CHAPTER 10

Time-Resolved Optical Responses of Solid

Time-resolved optical responses of solids are critically important as they offer a direct view into the fundamental, *ultrafast dynamics of energy flow and interactions* within materials, information that is inaccessible through conventional steady-state measurements. By utilizing extremely short laser pulses—ranging from femtosecond (10^{-15} s) to nanosecond (10^{-9} s) timescales—scientists can excite a solid and monitor, in real time, the subsequent relaxation processes and interactions among various components like electrons, holes, and phonons. This dynamic perspective is essential for developing a complete physical understanding of material behavior under nonequilibrium conditions. In time-resolved or pulsed optics, the concepts of the pulse and phase are fundamental to describing how light energy is packaged and travels in time and space, particularly for ultrashort laser systems.

10.1 PULSE AND PHASE

In time-resolved laser spectroscopy, a *pulse* refers to a short burst of electromagnetic radiation whose electric field varies rapidly in time. The electric field of such a pulse is commonly written as follows:

$$E(t) = A(t)\cos\left(\omega_0 t + \phi(t)\right) \tag{10.1}$$

where $A(t)$ is the slowly varying *envelope* that determines the intensity shape of the pulse, ω_0 is the carrier (optical) angular frequency, and $\phi(t)$ is the time-dependent phase, as shown in Figure 10.1. The measurable intensity of the pulse is proportional to the square of the envelope, $I(t) \propto |A(t)|^2$, meaning that the phase does not directly influence the intensity but becomes crucial for the temporal and spectral structure of ultrafast pulses. For many lasers, the envelope has a Gaussian profile, $A(t) = A_0 \exp(-t^2/ t\tau^2)$, where τ determines the full width at half maximum (FWHM) pulse duration $\Delta t = 2\sqrt{\ln 2}\ \tau$. The duration of the pulse ultimately sets the time resolution of a pump–probe experiment; for a pump pulse of width τ_p and a probe pulse of width τ_{pr}, the achievable temporal resolution is approximately

$$\Delta t_{\text{res}} = \sqrt{\tau_p^2 + \tau_{pr}^2} \tag{10.2}$$

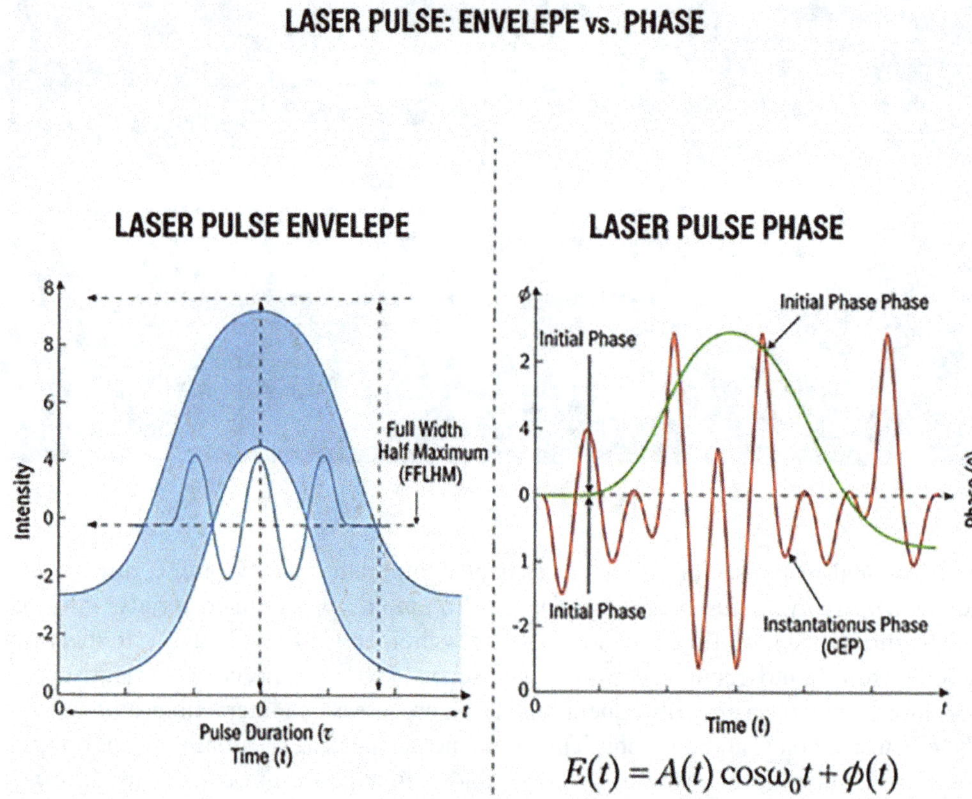

Figure 10.1 Diagram showing laser pulses and phases.
Diagram generated by AI assistance.

Along with the envelope, the *phase* of the pulse plays a central role in determining how different frequency components combine to form the field in time. The *carrier–envelope phase* (*CEP*) describes the offset between the peak of the carrier oscillation and the peak of the pulse envelope. This is captured by rewriting the electric field as follows:

$$E(t) = A(t)\cos(\omega_0 t + \phi_{CEP}) \tag{10.3}$$

where ϕ_{CEP} is constant for the pulse. CEP effects become important when pulses approach the few-cycle regime, allowing the relative alignment of the carrier and the envelope to influence the interaction with matter. However, in most time-resolved experiments using >20-fs pulses, CEP changes do not dominate the measurement.

More generally, ultrafast pulses are characterized by their *spectral phase*, which describes how the phase varies across the pulse's frequency components. The field in the frequency domain is

$$E(\omega) = \left|E(\omega)e\right|^{i\phi(\omega)}$$

and the spectral phase can be expanded around the central frequency ω_0 as follows:

$$\phi(\omega)=\phi_0+\phi_1(\omega-\omega_0)+\frac{1}{2}\phi_2(\omega-\omega_0)^2+\cdots . \tag{10.4}$$

Here, ϕ_1 corresponds to an overall time shift (or group delay), while ϕ_2 represents *group delay dispersion (GDD)* and is responsible for *pulse chirp*, the correlation of frequency with emission time. Chirp broadens the pulse: a pulse with initial transform-limited duration Δt_0 becomes longer when dispersion is present, according to

$$\Delta t=\sqrt{\Delta t_0^2+\left(\frac{4\ln 2\,\phi_2}{\Delta\omega}\right)^2} \tag{10.5}$$

where $\Delta\omega$ is the spectral bandwidth. Controlling spectral phase—through dispersion compensation or pulse shaping—is therefore essential to maintain short pulses and achieve high time resolution.

In pump–probe setups, the *relative delay* between pulses is controlled by adjusting the optical path length in one beam. A change in path ΔL produces a temporal delay:

$$\Delta t=\frac{\Delta L}{c}$$

allowing the experimenter to sample dynamical processes by observing how the probe pulse interacts with the sample after a controlled delay following excitation by the pump. Thus, both the *pulse* (its duration and envelope) and the *phase* (carrier–envelope and spectral) together determine the shape, timing, and effectiveness of ultrafast interactions in time-resolved laser experiments.

The *phase* refers to the position of the wave within its oscillation cycle at a given time and location. It is crucial because while the intensity envelope $A(t)$ determines *when* the energy arrives, and the phase

$$\Phi(t)=\omega_0 t+\phi(t) \tag{10.6}$$

determines *how* the wave interferes with itself or other waves.

The derivative of the phase with respect to time defines the *instantaneous frequency* $\omega(t)$ of the pulse:

$$\omega(t)=\frac{d\Phi(t)}{dt}=\omega_0+\frac{d\phi(t)}{dt} \tag{10.7}$$

In an ideal, transform-limited pulse, the envelope is symmetric, and the instantaneous frequency is constant ($\phi(t)=0$). However, real-world pulses often exhibit *chirp*, meaning the frequency changes within the pulse duration (e.g., the red part leads the blue part, or vice versa). Managing this phase manipulation is central to techniques like chirped pulse amplification (CPA) for generating the highest peak power lasers. The relationship between the pulse shape in time and its spectral bandwidth in frequency space is governed by the *Fourier Transform*, linking the temporal description to the spectral description of the pulse characteristics.

10.2 PUMP–PROBE TECHNIQUE

The *pump–probe technique* is a powerful experimental method in time-resolved spectroscopy used to study ultrafast phenomena in materials, such as the dynamics of excited charge carriers, phonons, and molecular vibrations, typically on femtosecond to nanosecond timescales. The core concept involves using two temporally delayed pulses: a strong *pump* pulse that excites the material into a nonequilibrium state and a weaker *probe* pulse that measures the subsequent changes in the material's optical properties as it relaxes back to equilibrium. The experimental setup is depicted in Figure 10.2.

The key to the pump–probe technique is precise control over the time delay (Δt) between the two pulses. The pump pulse hits the sample first at time $t = 0$. The probe pulse arrives at a variable time $t = \Delta t$ later. By measuring the change in the probe pulse's transmission or reflection as a function of this delay time, Δt, a time-evolution trace is mapped out.

The measured quantity is often the *differential transmission* ($\Delta T/T$), which quantifies how much the pump pulse altered the material's transparency:

$$\frac{\Delta T}{T}(\Delta t) = \frac{T_{\text{pump on}}(\Delta t) - T_{\text{pump off}}}{T_{\text{pump off}}}$$

where $T_{\text{pump on}}$ is the transmission of the probe when the pump is present and $T_{\text{pump off}}$ is the transmission of the probe alone. A positive $\frac{\Delta T}{T}$ signifies increased transparency (bleaching), while a negative value signifies increased absorption (induced absorption).

10.3 SEMICONDUCTOR RESPONSE

Pump–probe spectroscopy is an ultrafast optical technique used to study *carrier dynamics*, *excited state lifetimes*, and *band structure effects* in semiconductors.

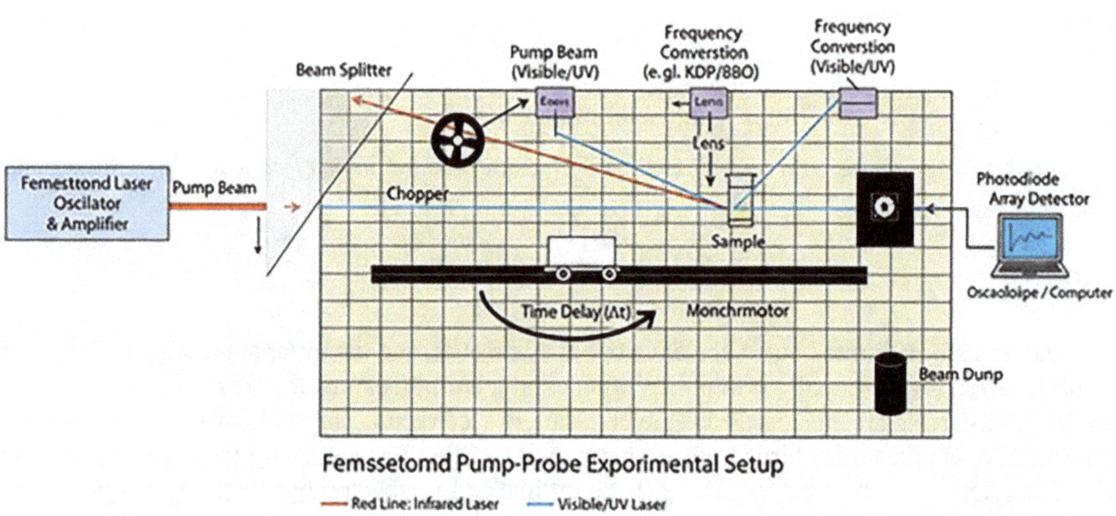

Figure 10.2 Pump–probe experimental apparatus.

Diagram generated by AI assistance.

A strong *pump* pulse excites carriers, and a weak *probe* pulse measures how the sample's transmission changes as a function of *delay time* τ. A femtosecond or picosecond laser pulse (*pump*) injects electrons and holes into the semiconductor. A second pulse (*probe*), time-delayed by τ, measures how the *transmission* $T(\tau)$ or *absorption* $\alpha(\tau)$ changes due to these carriers.

The probe pulses pass through the sample at varying delay times. The measured quantity is typically

$$\Delta T / T = \frac{T(\tau) - T_0}{T_0}$$

where T_0 is the transmission without the pump and $T(\tau)$ is the transmission with the pump at delay τ.

For a semiconductor with absorption coefficient $\alpha(\omega)$, the pump creates an excited carrier density:

$$N_0 = \frac{(1-R)F\alpha}{h\nu} \tag{10.8}$$

where R is the reflectivity, F is the pump fluence (J/m²), and $h\nu$ is the photon energy.

After excitation, changes in the probe transmission arise from the following:

a) State Filling (Pauli Blocking)

Electrons fill conduction band states; holes fill valence band states → absorption decreases.

$$\Delta\alpha_{\mathrm{SF}}(\tau) \propto -\left[f_c(\tau) + f_v(\tau)\right] \tag{10.9}$$

b) Bandgap Renormalization

Carrier–carrier interactions reduce the effective bandgap:

$$\Delta E_g(\tau) \propto -N(\tau)^{\frac{1}{3}} \tag{10.10}$$

This typically increases absorption near the band edge.

c) Free-Carrier Absorption (Drude Model)

One example, MoS2, is shown in Figure 10.3.

At long delays,

$$\Delta\alpha_{\mathrm{Drude}}(\omega) = \frac{Ne^2}{\epsilon_0\, cnm^*} \frac{1}{\omega^2 + \gamma^2} \tag{10.11}$$

- *Sharp rise* at $\tau = 0$ → carriers created by pump
- *Decay* → carrier recombination + trapping + cooling

The decay is often modeled as follows:

$$\Delta T / T(\tau) = \sum_i A_i e^{-\tau/\tau_i} \tag{10.12}$$

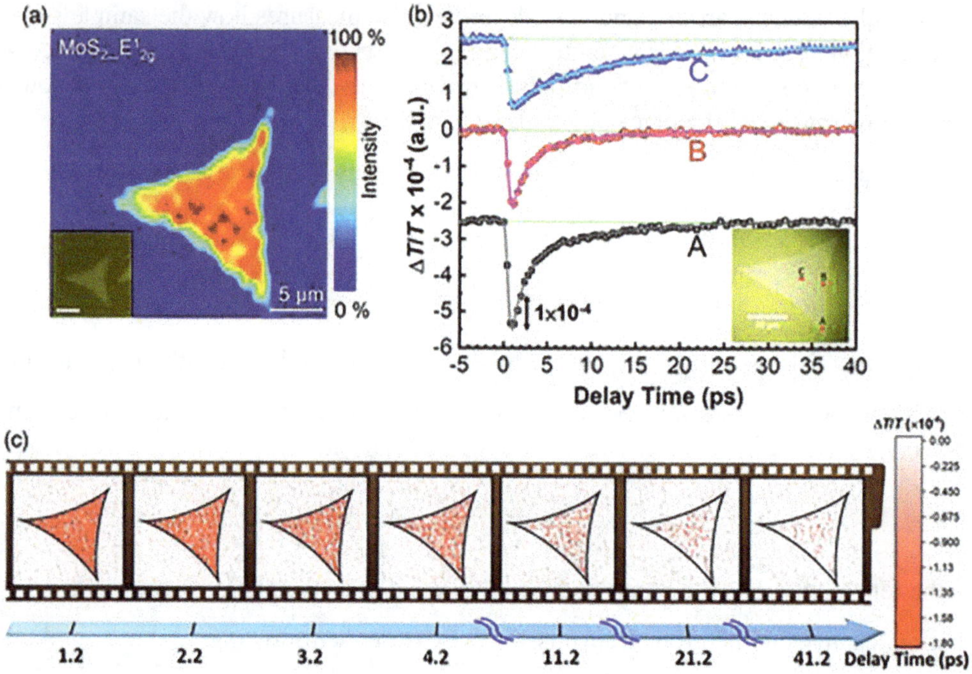

Figure 10.3 MoS2 pump–probe microscopy results.

Adv Photonics Res 3 (8), 2200046. OA.

IV

Optical Properties of Advanced Materials

V

Optical Properties of Advanced Materials

CHAPTER 11

Anisotropic Crystals

Anisotropic crystals play a crucial role in optics because their response to light depends on the direction of propagation and the polarization of the electric field, leading to phenomena that are impossible in isotropic materials. In such crystals, the dielectric function is a tensor rather than a scalar, so different polarization components of light experience different refractive indices, giving rise to birefringence, polarization-dependent absorption, and phase retardation. These properties enable precise control of optical polarization, which is essential for devices such as wave plates, polarizers, electro-optic modulators, nonlinear frequency converters, and polarization-sensitive detectors. More fundamentally, anisotropic optical behavior provides direct insight into crystal symmetry, lattice dynamics, and electronic structure, making polarization-resolved measurements a powerful tool in solid-state physics, materials science, and photonics. We will introduce a few of them in this chapter.

11.1 BIREFRINGENT CRYSTALS

We start with a plane–wave solution of the form

$$\mathbf{E}(\mathbf{r},t)=\mathbf{E}_0 e^{i(\mathbf{k}\cdot\mathbf{r}-\omega t)}$$

which reduces to an eigenvalue problem relating $\mathbf{k}$, ω, and $\mathbf{E}_0$:

$$\left[k^2\mathbf{I}-\mathbf{kk}-\frac{\omega^2}{c^2}\varepsilon\right]\mathbf{E}_0=0 \tag{11.1}$$

Nontrivial solutions exist only if the determinant of the bracketed matrix vanishes, which yields the *dispersion relations* for electromagnetic waves in the crystal.

For a *uniaxial birefringent crystal*, the dielectric tensor in its principal axes is diagonal:

$$\varepsilon=\begin{pmatrix}\varepsilon_o & 0 & 0\\ 0 & \varepsilon_o & 0\\ 0 & 0 & \varepsilon_e\end{pmatrix}$$

with refractive indices $n_o = \sqrt{\varepsilon_o}$ and $n_e = \sqrt{\varepsilon_e}$

Solving the eigenvalue equation shows that there are two allowed wave solutions for a given propagation direction. One solution has its electric field perpendicular to the plane containing the wave vector and the optical axis; this is the *ordinary wave*, with dispersion relation:

$$k^2 = \frac{\omega^2}{c^2}\varepsilon_o \Rightarrow k = \frac{n_o\omega}{c}$$

independent of direction. The second solution, the *extraordinary wave*, has its electric field in the plane defined by the wave vector and the optical axis, and its dispersion relation depends on the propagation angle θ:

$$\frac{k^2}{\omega^2 / c^2} = \frac{\varepsilon_o\varepsilon_e}{\varepsilon_o \sin^2\theta + \varepsilon_e \cos^2\theta} \tag{11.2}$$

This angular dependence of the wave vector magnitude leads to the direction-dependent refractive index $n_e(\theta)$.

Because the ordinary and extraordinary waves have different wave vectors, they propagate with different phase velocities $v_p = \omega/k$, producing a phase difference as the light travels through a crystal of thickness d. From the Maxwell-based dispersion relations, the accumulated phase retardation is

$$\Delta\phi = (k_e - k_o)d = \frac{2\pi}{\lambda}(n_e - n_o)d$$

which explains polarization changes and the operation of wave plates directly from electromagnetic theory.

Thus, starting from Maxwell's equations and the anisotropic constitutive relation $\mathbf{D} = \varepsilon_0\boldsymbol{\varepsilon}\mathbf{E}$, one finds that electromagnetic waves in birefringent crystals naturally split into two orthogonally polarized eigenmodes with different dispersion relations. This Maxwell equation–based treatment provides the fundamental explanation for double refraction, phase retardation, and polarization control in birefringent crystals.

11.2 NONLINEAR CRYSTALS

An optical frequency doubler, or second-harmonic generator, uses a nonlinear crystal with a second-order susceptibility to convert incoming light of frequency ω into new light at 2ω. When a linearly polarized wave enters the crystal, the electric field drives a nonlinear polarization term, which oscillates at twice the original frequency and therefore radiates light at 2ω. Efficient generation requires *phase matching*, meaning the newly created second-harmonic wave must stay in phase with the nonlinear polarization as they both propagate; this is achieved by exploiting birefringence or, in quasi-phase–matched materials like Periodically Poled Lithium Niobate (PPLN), by periodically flipping the nonlinear coefficient. When phase matching is satisfied, the second-harmonic field builds coherently along the crystal, giving an output beam that contains both the original ω light and the newly generated 2ω light, whose intensity grows roughly as the square of the input intensity and crystal length.

11.2.1 Nonlinear Polarization

The origin of second-harmonic light comes from the second-order nonlinear term in the material's polarization:

$$P(t) = \varepsilon_0 \left(\chi^{(1)} E + \chi^{(2)} E^2 + \ldots \right) \tag{11.3}$$

The part generating SHG is

$$P_{2\omega} = \varepsilon_0 \chi^{(2)} E_0^2 \cos(2\omega t) \tag{11.4}$$

11.2.2 Phase Matching Condition

For efficient conversion, the generated 2ω wave must stay in phase with the nonlinear polarization:

$$k_{2\omega} = 2k_\omega$$

In terms of refractive index,

$$n_{2\omega} = n_\omega$$

(or practically matched using birefringence or quasi-phase matching).

11.2.3 Wave Vector Mismatch

The degree of mismatch is

$$\Delta k = k_{2\omega} - 2k_\omega$$

This determines how efficiently SHG builds up.

11.2.4 Conversion Efficiency (Simplified)

For a perfect phase-matched crystal,

$$I_{2\omega} \propto I_\omega^2 L^2 \, d_{\text{eff}}^2 \tag{11.5}$$

where I_ω is the input intensity, L is the crystal length, and d_{eff} is the effective nonlinear coefficient (related to χ^2).

11.3 NOTCH FILTERS

An optical notch filter, also known as a band-stop or band-rejection filter, is designed to selectively block a very specific, narrow range of wavelengths while allowing light outside of this rejected band (both shorter and longer wavelengths) to pass through with minimal attenuation. The primary mechanism for achieving this precise filtering action is through optical interference, although some broader-band filters rely on absorption principles. This specialized function is crucial in applications where a single, intense wavelength of light must be precisely removed from a complex light signal.

The mechanism behind most high-performance notch filters involves multilayer dielectric (dichroic) coatings. The filter consists of alternating, extremely thin layers of materials with different refractive indices, with the thickness of these layers precisely engineered to correspond to the target wavelength intended for blocking. When light within this specific target range enters the filter, reflections at each material interface are timed to align perfectly, causing destructive interference that cancels out the light wave. This action highly reflects the unwanted band of wavelengths away from the transmission path, while wavelengths outside the target range pass through the layers efficiently due to constructive interference.

Notch filters are indispensable in scientific and industrial fields, particularly when a powerful light source needs to be removed to observe weaker signals nearby in the spectrum. A common application is in Raman spectroscopy, where the intense laser light (Rayleigh scattering) must be blocked to enable the detection of the much weaker, adjacent Raman signals that provide chemical data. They are also vital for laser safety applications, providing protective eyewear or instrument filters designed to block hazardous specific laser wavelengths while allowing normal vision or operation to occur with high transmission efficiency of nonhazardous light.

Here is a diagram illustrating how an optical notch filter blocks specific wavelength photons:

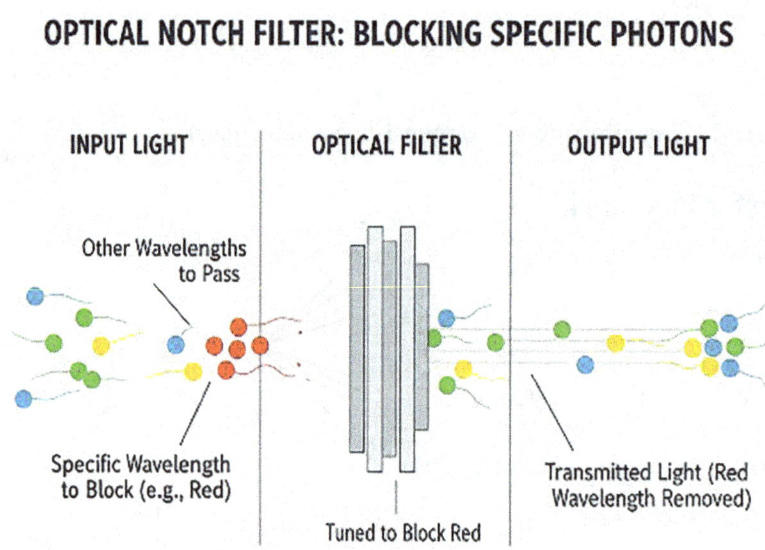

Diagram generated by AI assistance.

11.3.1 Interference-Based Notch Filter (Optical Principle)

Many optical notch filters operate by *destructive interference*. The incident light is split into two paths: one is transmitted directly, while the other acquires a frequency-dependent phase shift and then recombines with the first. The total transmitted field is

$$E_{\text{out}}(\omega) = E_0\left[1 + e^{i\phi(\omega)}\right] \tag{11.6}$$

where $\phi(\omega)$ is the relative phase difference between the two paths. The transmission intensity becomes

$$T(\omega) \propto \left|1 + e^{i\phi(\omega)}\right|^2 = 2\left[1 + \cos\phi(\omega)\right] \tag{11.7}$$

At frequencies where $\phi(\omega_0) = \pi$, the cosine term equals -1, and destructive interference occurs, yielding

$$T(\omega_0) = 0$$

which produces the notch. For frequencies away from ω_0, the phase mismatch is not exactly π, so transmission is high.

11.4 LOW-PASS FILTER

An optical long-pass filter (LPF) functions by selectively transmitting light wavelengths that are longer than a specific "cut-on" wavelength while blocking all shorter wavelengths, a process achieved through two primary physical principles. Absorption-based filters utilize colored materials that absorb the energy of shorter wavelengths, converting it to heat and allowing longer wavelengths like red or infrared light to pass through. Conversely, interference (dichroic) filters use multiple, precisely engineered thin layers that employ optical interference: shorter wavelengths are reflected away through destructive interference, while longer wavelengths pass through due to constructive interference, offering a much sharper, more precise transition between blocked and transmitted light. This mechanism makes LPFs essential in applications such as fluorescence microscopy, where they isolate longer-wavelength emission signals from shorter-wavelength excitation light, and in photography to facilitate specialized infrared imaging by blocking visible light.

Here is a diagram illustrating how an optical LPF blocks specific wavelength photons:

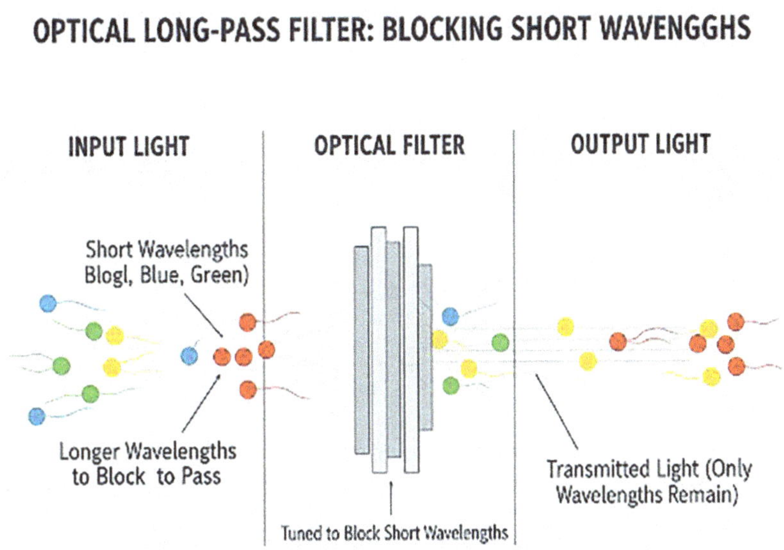

Diagram generated by AI assistance.

11.4.1 First-Order Low-Pass Filter

The simplest and most common low-pass filter is a *first-order filter*, whose transfer function is

$$H(\omega)=\frac{1}{1+\dfrac{i\omega}{\omega_c}} \tag{11.8}$$

The corresponding transmission intensity is

$$T(\omega)=\frac{1}{1+(\omega/\omega_c)^2} \tag{11.9}$$

At low frequencies ($\omega \ll \omega_c$), the transmission approaches unity, meaning the signal passes through nearly unchanged. At the cutoff frequency $\omega = \omega_c$, the transmission drops to

$$T(\omega_c)=\frac{1}{2}$$

which corresponds to a –3 dB point. For frequencies much higher than the cutoff ($\omega \gg \omega_c$), the transmission decreases as $T(\omega) \propto \omega^{-2}$, strongly suppressing high-frequency components.

In addition to amplitude attenuation, a low-pass filter introduces a *frequency-dependent phase shift*:

$$\phi(\omega)=-\tan^{-1}\left(\frac{\omega}{\omega_c}\right) \tag{11.10}$$

This phase response leads to a *group delay*:

$$\tau_g(\omega)=-\frac{d\phi}{d\omega}=\frac{1}{\omega_c\left[1+(\omega/\omega_c)^2\right]} \tag{11.11}$$

which means that different frequency components of a pulse experience different delays, potentially broadening time-domain signals.

More complex low-pass filters (higher-order) are formed by cascading multiple stages. For an n-th order low-pass filter, the transmission intensity becomes

$$T(\omega)=\frac{1}{1+(\omega/\omega_c)^{2n}} \tag{11.12}$$

which yields a much sharper roll-off beyond the cutoff frequency. Such filters are used when a strong rejection of high frequencies is required while maintaining high transmission at low frequencies.

CHAPTER 12

Magnetic Materials

12.1 MAGNETO-OPTICAL KEFF EFFECT

The *optical Kerr effect (OKE)* is a manifestation of third-order optical nonlinearity where a material's refractive index changes in direct proportion to the intensity of the light wave passing through it. Unlike the DC Kerr effect, which uses an external static field, the OKE is self-induced by the intense electric field of the light beam itself.

12.1.1 Mechanism of the Optical Kerr Effect

The physical mechanism typically involves the interaction of the intense light field with the electron cloud or molecular structure of the material:

1. **Electronic Polarization:** The electric field of the light distorts the electron orbitals of the atoms/molecules.
2. **Molecular Reorientation:** In certain liquids, the electric field exerts a torque on polar molecules, causing them to align with the field. This alignment creates a local anisotropy and temporary birefringence.

In nonlinear optics, the polarization **P** induced in a medium by an electric field **E** is expanded as a power series:

$$\mathbf{P} = \varepsilon_0 \left(\chi^{(1)}\mathbf{E} + \chi^{(2)}\mathbf{E}^2 + \chi^{(3)}\mathbf{E}^3 + \cdots \right) \tag{12.1}$$

For centrosymmetric media (such as glasses and liquids), the second-order susceptibility $\chi^{(2)}$ vanishes, and the lowest nonlinear contribution is the third-order term $\chi^{(3)}$. The OKE arises from this $\chi^{(3)}$ term and represents an *intensity-dependent modification of the linear polarization.*

For a monochromatic optical field, the third-order nonlinear polarization leads to a refractive index that depends on the light intensity I:

$$n = n_0 + n_2 I \tag{12.2}$$

where n_0 is the linear refractive index and n_2 is the *Kerr coefficient*, directly related to $\chi^{(3)}$ by

$$n_2 = \frac{3}{4n_0^2\varepsilon_0 c}\operatorname{Re}\left[\chi^{(3)}\right] \tag{12.3}$$

This equation shows that higher light intensity increases (or decreases, depending on the sign of n_2) the refractive index of the medium.

Let us next examine the interaction between linearly polarized light and magnetic materials. A famous phenomenon is known as the Magneto-optic Kerr effect (*MOKE*). There are three situations depending on the magnetization orientation of the magnetic materials, as summarized in Figure 12.1:

MOKE is observed because the reflected light from the magnetic material changes its polarization and intensity after the interaction, resulting from the anisotropic dielectric permittivity in such magnetic materials. We often quantify MOKE with the Kerr angle θ_k or Kerr ellipticity η_k. The Kerr angle is the angle of rotation measured after the linearly polarized light is reflected from the sample, usually less than 1 degree for many materials. Kerr ellipticity is the ratio of the two axes of the elliptically polarized light intensity measured after the linearly polarized light is reflected from the sample, usually also less than 1 degree.

When linearly polarized light is incident on a magnetized medium, the dielectric response becomes anisotropic and is described by a *magneto-optical dielectric tensor*. For a ferromagnetic material with magnetization **M**, the dielectric tensor can be written as

$$\varepsilon = \begin{pmatrix} \varepsilon & -iQM_z & iQM_y \\ iQM_z & \varepsilon & -iQM_x \\ -iQM_y & iQM_x & \varepsilon \end{pmatrix} \tag{12.4}$$

Name	(a) Polar	(b) Longitudinal	(c) Transverse
Geometry	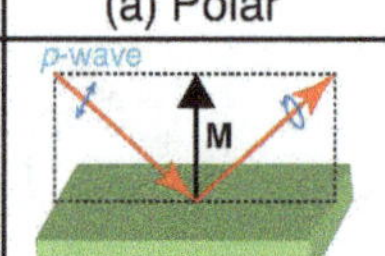	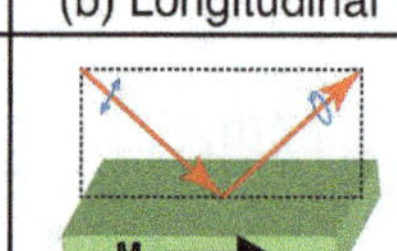	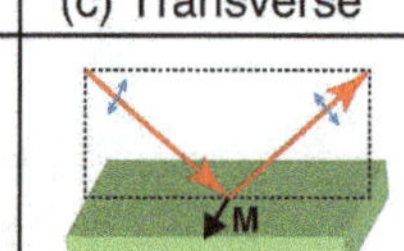
Detection	Out-of-plane	In-plane	In-plane
Polarization variation	Rotation ellipticity		None
Measurement	Polarization analysis		Intensity measurement

Figure 12.1 MOKE setups.

Appl. Sci. 7, 662 (2017). OA.

where ε is the diagonal (nonmagnetic) dielectric constant and Q is the *magneto-optical constant*. The off-diagonal elements, proportional to **M**, are responsible for the Kerr effect.

Due to these off-diagonal terms, the reflection coefficients for right- and left-circularly polarized light become different. For polar MOKE (magnetization perpendicular to the surface), the complex Kerr angle is defined as

$$\Phi_K = \theta_K + i\varepsilon_K = \frac{r_{ps}}{r_{pp}} \tag{12.5}$$

where r_{pp} is the Fresnel reflection coefficient for p-polarized light and r_{ps} is the magnetization-induced cross-polarized reflection coefficient. The real part θ_K represents the *Kerr rotation* (rotation of polarization), and the imaginary part ε_K represents the *Kerr ellipticity*.

In a typical MOKE experiment, the change in *reflected intensity* is detected after an analyzer (polarizer). If the incident light is linearly polarized and the analyzer is set at an angle α relative to the incident polarization, the reflected intensity is

$$I = I_0 \sin^2(\alpha + \theta_K) \tag{12.6}$$

where I_0 is the reflected intensity without magnetization. For small Kerr rotations ($\theta_K \ll 1$), this expression can be expanded to first order:

$$I \approx I_0 \left[\sin^2 \alpha + 2\theta_K \sin\alpha \cos\alpha\right] \tag{12.7}$$

The *magnetization-dependent intensity change* is therefore

$$\Delta I \propto \theta_K \propto M$$

showing that the detected signal is directly proportional to the sample magnetization.

There are two significant MOKE results worth mentioning here: MOKE microscopy reported in Figure 12.2, and MOKE hysteresis extracted in Figure 12.3. They both reflect the magnetic domain behavior with magneto-optical responses.

To construct MOKE microscopy, we could include one objective in the incident beam path shown in the geometry illustration in Figure 9.4 before the incident light reaches the sample surface. In this case, once the Kerr angle θ_k is obtained for the sample, raster scanning can be performed to produce a microscopic image, and the magnetic domains could be identified with MOKE contrasts as depicted in Figure 12.2. It is worth mentioning that the contrast for normal MOKE signals at different domains is usually small, and research efforts are being made to improve such contrast. For example, an extreme antireflection structure is shown in Figure 12.2.

Now, if MOKE could identify magnetic domains and these domains respond to magnetic fields, it is natural to apply MOKE to observe hysteresis as well. Figure 12.3 shows the hysteresis loop of a thin film Ta (2 nm)/Gd21(Fe90Co10)79 (20 nm)/Ta (10 nm)/SiO_2.

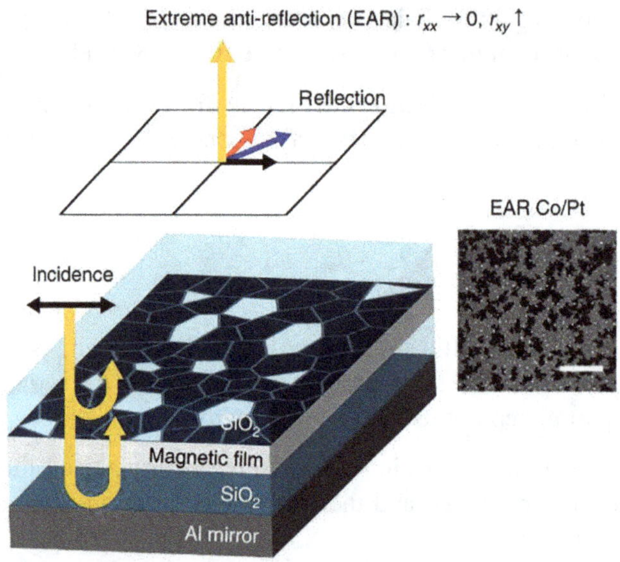

Figure 12.2 MOKE microscopy. Scale bar, 50 μm.

Nat. Commun. 11, 5937 (2020). OA.

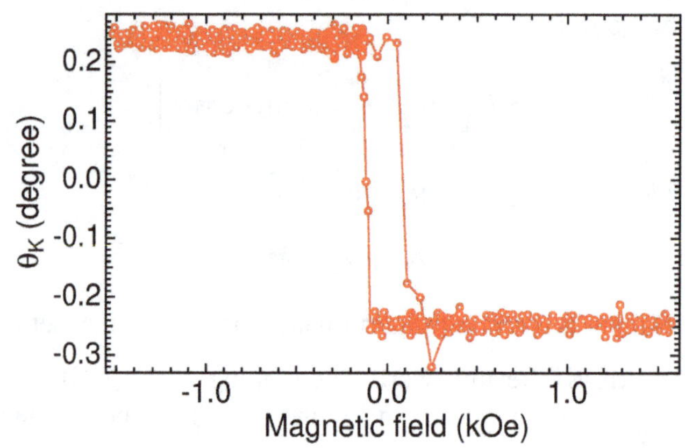

Figure 12.3 MOKE hysteresis.

Appl. Sci. 7, 662 (2017). OA.

12.2 FARADAY EFFECT

The *Faraday effect*, also known as *Faraday rotation*, is a fundamental phenomenon in *magneto-optics*, where the polarization plane of light rotates as it passes through a material under the influence of a magnetic field aligned in the direction of light propagation. It was discovered by Michael Faraday in 1845 and is one of the first observed interactions between light and magnetic fields. This effect is a manifestation of the interaction between the electromagnetic field of the light and the magnetic properties of the medium, and it has applications in optical isolators, modulators, and magnetic field sensors.

When linearly polarized light propagates through a medium of length L in the presence of a magnetic field B along the propagation direction, the plane of polarization rotates by an angle θ given by

$$\theta = VBL \tag{12.8}$$

where V is the *Verdet constant* of the material, which depends on the wavelength of light and the material's properties; B is the magnetic flux density along the light propagation direction; and L is the path length of the light in the material. The Verdet constant is typically measured in radians per tesla per meter ($rad/(T{\cdot}m)$).

The Verdet constant measures how strongly a material interacts with the magnetic field to rotate light. A larger Verdet constant means that the material causes a larger rotation of the plane of polarization for a given magnetic field and length.

It depends on the following:

- Wavelength of light (λ): Shorter wavelengths usually give larger Verdet constants.
- Temperature (T): For some materials, the Verdet constant decreases with increasing temperature.
- Material properties: For example, heavy glass or certain crystals like terbium gallium garnet (TGG) have very high Verdet constants.

Typical Values

- Common optical glasses: 0.01 – 0.1 $rad/(T{\cdot}m)$
- TGG: ~40 $rad/(T{\cdot}m)$ at 632.8 nm (He-Ne laser wavelength)

The physical origin of the Faraday effect can be understood in terms of circular birefringence as shown in Figure 12.4. In the presence of a magnetic field, the refractive indices for right- and left-circularly polarized light (n_+ and n_-) become different. A linearly polarized wave can be decomposed into these two circular components. As they travel through the medium, they acquire a phase difference:

$$\Delta\phi = \frac{2\pi L}{\lambda}\left(n_+ - n_-\right) \tag{12.9}$$

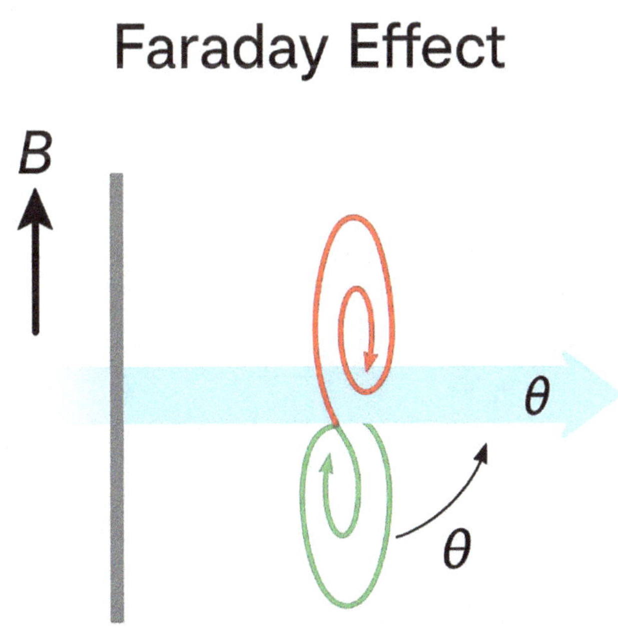

Figure12.4 Faraday effect in combination with left- and right-handed light.

Diagram generated by AI assistance.

This phase difference causes the rotation of the polarization plane:

$$\theta = \frac{\Delta\phi}{2} = \frac{\pi L}{\lambda}(n_+ - n_-) \tag{12.10}$$

where λ is the wavelength of light in vacuum. The difference in refractive indices arises because the magnetic field induces circular dichroism in the medium, which interacts differently with left- and right-circularly polarized components.

The Faraday effect is *nonreciprocal*, meaning that if the light is reflected back through the medium, the polarization rotation does *not cancel* but instead doubles. This property is essential in devices such as optical isolators, which prevent reflected light from destabilizing lasers. In essence, the Faraday effect bridges magnetism and optics, offering a direct way to manipulate light polarization using magnetic fields.

If the transmitted light passes through a polarizer (analyzer) oriented at an angle α relative to the incident polarization, the detected intensity is

$$I = I_0 \cos^2(\alpha - \theta_F) \tag{12.11}$$

where I_0 is the intensity before the analyzer. For small rotation angles ($\theta_F \ll 1$), this can be approximated as follows:

$$I \approx I_0\left[\cos^2\alpha + 2\theta_F \cos\alpha \sin\alpha\right] \tag{12.12}$$

The *magnetization-dependent intensity change* is then

$$\Delta I \propto \theta_F \propto M_z,$$

meaning that the transmitted intensity varies linearly with the magnetization component along the light propagation direction.

12.3 MAGNETO PHOTOLUMINESCENCE

Next, let us examine the optical response involving the angular momentum of circularly polarized light. Specifically, we will discuss photoluminescence (PL) circular polarization defined as follows:

$$P = \frac{I_+ - I_-}{I_+ + I_-} \tag{12.13}$$

where I_+ and I_- are the intensities of σ_+ and σ_- circularly polarized PL components as defined in Section 9.1, respectively. Before analyzing this phenomenon further, we should introduce the basic concept of PL. PL is defined as light emission from the sample after it has absorbed incident photons. You may ask: what kind of samples can have PL? This requires us to examine the band structure more closely. Here, we are going to take one two-dimensional semiconductor as an example: WS_2. The band structure of this monolayer (less than 1 nm thick) WS_2 is calculated in Figure 12.5:

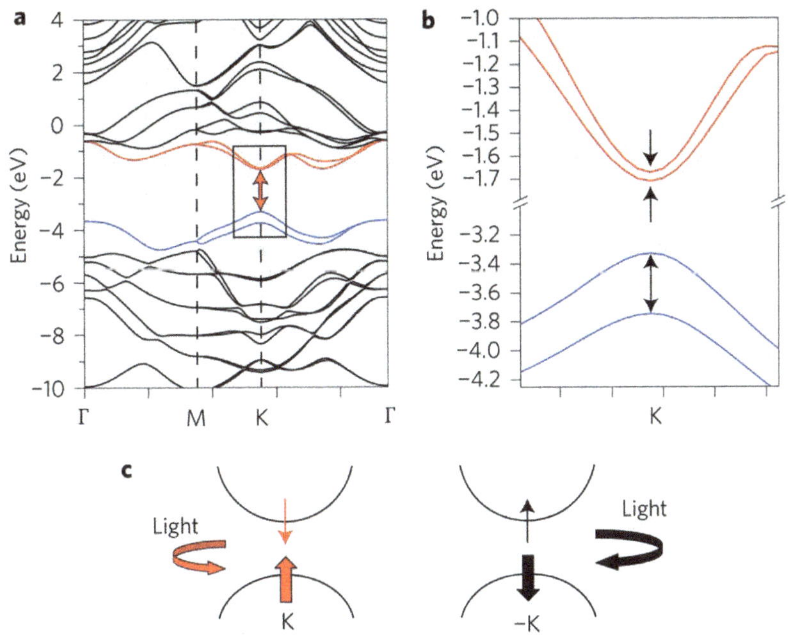

Figure 12.5 Band structure of WS2 monolayer.

Nat. Nanotech. 10, 603–607 (2015). Au owned.

Unlike the band structure of Si, where the valence band (last completely filled band, imagine completely filled atomic shells) maximum (*VBM*) is at the *Γ* point and the conduction band (the first unfilled band, imagine partially filled atomic shells) minimum (*CBM*) is at the *X* point, for the WS_2 monolayer (yes, I emphasize this not only because it is 2D but also because the band structure of bilayer WS_2 is different), both the VBM and CBM are at the *K* point. This distinguishes the two semiconductors as indirect bandgap semiconductors (Si) and direct bandgap semiconductors (WS_2 monolayer). Why is this critical for PL discussion? Basically, only direct bandgap semiconductors can have PL. The reason is as follows: for any visible light emission, the radiation wavelength is between 400 and 750 nm, which means the corresponding wave number *k* is extremely small in comparison with lattice constants in subnanometers. Therefore, following the momentum conservation law (again, why we use *k* space), only energy transitions that have almost no wave number change are allowed. It means that the VBM and CBM will have to be at the same *k* space coordinates, resulting in a direct bandgap. Notice the splitting at VBM and CBM at *K* shown in Figure 12.5. It is due to spin–orbit interactions, and this splitting leads to a certain orientation-favored spin (e.g., spin down in *K*) staying at the lower splitting level. When this happens, only the incident light with σ_+ polarization can excite the electrons from VBM to this CBM splitting level. After electrons in this *K*-valley relax back to VBM and generate PL, naturally, only one PL circular polarization occurs, and ideally, $P = 100\%$. Due to the crystal inversion symmetry breaking of the 2H-phase WS_2 monolayer, the *K*-valley and (–*K*)-valley have different orientation-favored spins in the CBM, hence the spin-valley locking. Therefore, for the (–*K*)-valley, ideally, the PL circular polarization would be –100%. Figure 12.6 shows the *k* space of WS_2 for our convenience.

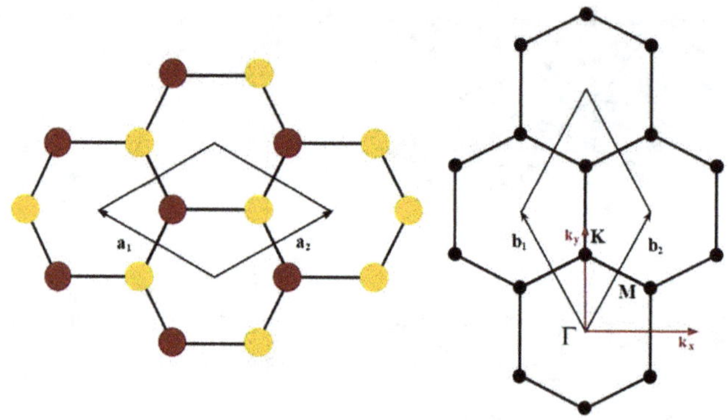

Figure 12.6 Reciprocal lattice of WS2.

Mater. Adv. 3, 5845–5856 (2022).

CHAPTER 13

Superconductors

In a superconductor, the carriers responsible for transporting electric current are known as *Cooper pairs*. Unlike in ordinary conductors, where individual electrons scatter and create resistance, electrons in a superconducting state form paired states through interactions with the crystal lattice. These Cooper pairs move in a coordinated, wave-like manner that prevents scattering, allowing electric current to flow without resistance. Their collective behavior is what enables the defining properties of superconductors, including zero electrical resistance and the expulsion of magnetic fields.

13.1 LONDON MODEL

The London model is an early theory of superconductivity proposed by Fritz and Heinz London in 1935 to explain how superconductors interact with electromagnetic fields. It assumes that superconducting charge carriers—now known to be Cooper pairs—move without resistance and respond collectively to electric and magnetic fields. Even though the London model does not describe the microscopic origin of superconductivity, it successfully explains key electromagnetic properties, most importantly the Meissner effect, where superconductors expel magnetic fields as they transition into the superconducting state. By introducing simple equations governing current and magnetic field behavior, the London model laid the foundation for more advanced theories.

The London equations are two fundamental relations that describe how superconducting current behaves under electric and magnetic fields. The first London equation states that a superconducting current accelerates continuously in response to an electric field, reflecting the absence of electrical resistance. The second equation relates the curl of the supercurrent to the magnetic field, and when combined with Maxwell's equations, it predicts that magnetic fields cannot deeply penetrate a superconductor. Together, these two equations capture the essential electromagnetic behavior of superconductors in the simplest possible form.

13.1.1 First London Equation

$$\frac{\partial \mathbf{J}_s}{\partial t} = \frac{n_s e^2}{m}\mathbf{E} \tag{13.1}$$

where $\mathbf{J}_s$ is the supercurrent density, n_s is the density of superconducting electrons, e is the electron charge, m is the electron mass, and $\mathbf{E}$ is the electric field.

In a superconductor, the current responds without resistance to an applied electric field, leading to *persistent currents.*

13.1.2 Second London Equation

$$\nabla \times \mathbf{J}_s = -\frac{n_s e^2}{m}\mathbf{B} \tag{13.2}$$

where $\mathbf{B}$ is the magnetic field.

This equation explains the *Meissner effect*, showing that the magnetic field decays exponentially inside a superconductor.

From the second London equation, one can derive that magnetic fields decay exponentially inside a superconductor, rather than simply being blocked at the surface. This decay occurs over a characteristic length called the *London penetration depth*, which depends on the density and mass of the superconducting carriers. The penetration depth is typically on the order of tens of nanometers and represents how far a magnetic field can enter before being suppressed. This mathematical result explains the Meissner effect and is one of the most important insights produced by the London theory. From Maxwell's equation,

$$\nabla \times \mathbf{B} = \mu_0 \mathbf{J}_s$$

Substituting the second London equation,

$$\nabla \times (\nabla \times \mathbf{B}) = -\frac{n_s e^2}{m}\mu_0 \mathbf{B}$$

Using the vector identity $\nabla \times (\nabla \times \mathbf{B}) = \nabla(\nabla \cdot \mathbf{B}) - \nabla^2 \mathbf{B}$ and $\nabla \cdot \mathbf{B} = 0$,

$$-\nabla^2 \mathbf{B} = -\frac{n_s e^2 \mu_0}{m}\mathbf{B}$$

$$\nabla^2 \mathbf{B} = \frac{1}{\lambda_L^2}\mathbf{B}$$

where λ_L is the *London penetration depth*:

$$\lambda_L = \sqrt{\frac{m}{\mu_0 n_s e^2}} \tag{13.3}$$

For a magnetic field entering the superconductor along the *x*-direction,

$$B(x) = B_0 e^{-x/\lambda_L}$$

This shows the exponential decay of the magnetic field inside the superconductor. Typically, λ_L is on the order of tens to hundreds of nanometers.

The London model provides a phenomenological way to describe this for *superconductors*, particularly at frequencies below the superconducting gap.

The *first London equation* in the time domain is as follows:

$$\frac{\partial \mathbf{J}_s}{\partial t} = \frac{n_s e^2}{m}\mathbf{E}$$

Taking a *Fourier transform* to the frequency domain ($\mathbf{E}(t) \rightarrow \mathbf{E}(\omega)$),

$$i\omega \mathbf{J}_s(\omega) = \frac{n_s e^2}{m}\mathbf{E}(\omega)$$

$$\mathbf{J}_s(\omega) = \frac{n_s e^2}{m}\frac{1}{i\omega}\mathbf{E}(\omega) \tag{13.4}$$

By the definition of conductivity,

$$\mathbf{J}(\omega) = \sigma(\omega)\mathbf{E}(\omega)$$

So the *complex conductivity of the superconductor* in the London model is

$$\sigma_s(\omega) = \frac{n_s e^2}{m}\frac{1}{i\omega} = i\frac{n_s e^2}{m\omega} \tag{13.5}$$

13.1.3 Implications for Optical Properties

a) ***Real and Imaginary Parts***

$$\sigma_s(\omega) = \sigma_1(\omega) + i\sigma_2(\omega)$$

- Here, $\sigma_1(\omega) = 0 \rightarrow$ *no absorption* because supercurrent flows without resistance.
- $\sigma_2(\omega) = \frac{n_s e^2}{m\omega} \rightarrow$ *inductive response.*

The superconductor does not dissipate energy at low frequencies; it behaves like a *perfect inductor*.

b) ***Dielectric Function***

The dielectric function is related to conductivity by

$$\varepsilon(\omega) = \varepsilon_\infty + \frac{i\sigma(\omega)}{\varepsilon_0\omega}$$

Substituting $\sigma_s(\omega)$,

$$\varepsilon(\omega)=\varepsilon_\infty+\frac{i}{\varepsilon_0\omega}\left(i\frac{n_s e^2}{m\omega}\right)=\varepsilon_\infty+\frac{n_s e^2}{\varepsilon_0 m\omega^2} \tag{13.6}$$

Define the *plasma frequency* of superconducting electrons:

$$\omega_{ps}^2=\frac{n_s e^2}{\varepsilon_0 m}$$

Then,

$$\varepsilon(\omega)=\varepsilon_\infty+\frac{\omega_{ps}^2}{\omega^2} \tag{13.7}$$

Key point: The dielectric constant diverges as $\omega \to 0$, reflecting *perfect screening of static fields* (Meissner effect).

13.2 EXCITATIONS

Optical excitation in superconductors occurs when photons interact with the material, creating excitations above the superconducting ground state. Photons with energy $h\nu > 2\Delta$ can break Cooper pairs, producing quasiparticles, while photons with $h\nu < 2\Delta$ interact with the superfluid condensate. The optical conductivity is given by $\sigma(\omega) = \sigma_1(\omega) + i\sigma_2(\omega)$, where σ_1 is absorption from quasiparticles and σ_2 is the superconducting response. The London equations

$$\nabla\times\mathbf{J}_s=-\frac{n_s e^2}{m}\mathbf{B}$$

and

$$\frac{d\mathbf{J}_s}{dt}=\frac{n_s e^2}{m}\mathbf{E}$$

describe the superfluid's response, and the London penetration depth

$$\lambda_L=\sqrt{\frac{m}{\mu_0 n_s e^2}}$$

sets the field penetration. Optical measurements allow determination of the superconducting gap Δ, superfluid density n_s, and quasiparticle dynamics, making them essential for studying conventional and high-temperature superconductors, as shown in Figure 13.1.

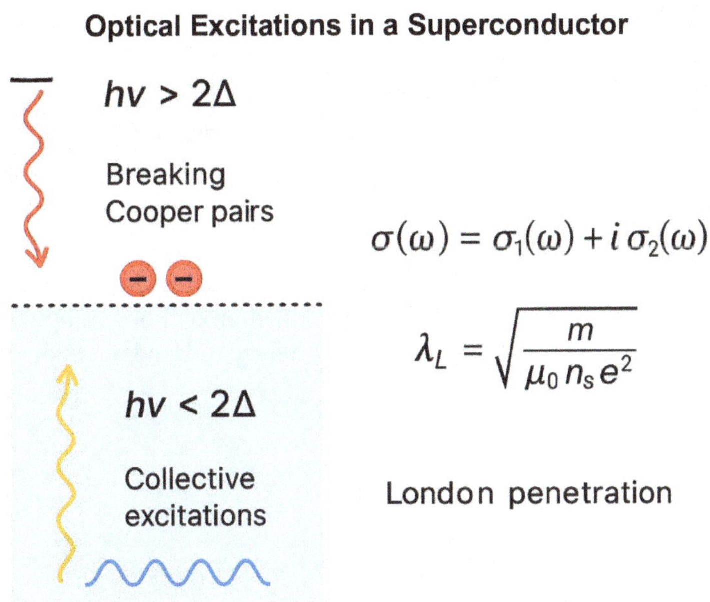

Figure 13.1 Excitation energy in comparison with superconducting gap.

Diagram generated by AI assistance.

13.3 OPTICAL CONDUCTIVITY OF SUPERCONDUCTOR

The London model is important for understanding the optical response of superconductors because it provides the simplest theoretical framework that predicts how superconductors interact with electromagnetic waves—especially at low frequencies. Its equations explain how superconducting currents respond to electric and magnetic fields, and this directly shapes how a superconductor absorbs or reflects light.

Here are the key reasons:

a) Predicts Zero-Frequency (Low-Energy) Optical Behavior

The London equations show that superconductors have infinite DC conductivity. In optical terms, this means that at very low frequencies (microwave to THz range), the optical conductivity has a strong response from the superconducting condensate. This produces a sharp feature known as the delta function at zero frequency, which is a hallmark of superconductivity in optical spectra.

b) Explains the Meissner Effect and Reflection of Electromagnetic Waves

Because the London model predicts an exponential decay of magnetic fields, it implies that electromagnetic waves cannot penetrate deeply into a superconductor. This gives rise to extremely high reflectivity at low frequencies, which is directly observed in optical experiments. The London penetration depth determines how much of the field actually enters and thus influences the measured optical response.

c) Connects Optical Conductivity to Superfluid Density

Optical measurements can determine how much of the electron population participates in superconductivity.

In the London model,

$$\lambda_L^{-2} \propto n_s$$

Because the penetration depth comes from the London equations, measuring optical reflectivity and transmission allows researchers to extract the superfluid density, an essential parameter in superconductors.

d) Provides the Foundation for More Advanced Optical Theories

The London model serves as the starting point for more complete descriptions, such as the two-fluid model and Mattis–Bardeen theory, which give detailed predictions of optical conductivity across different frequency ranges. These later models all reduce to London behavior at low frequencies.

CHAPTER 14

Quantum Structures

14.1 LIGHT ABSORPTION BY CONFINED ELECTRONS IN QUANTUM WELLS

We start by discussing standing waves in quantum wells:

$$\vec{E}(z)=\vec{E}_{xy}\,E_0\,cos(qz)\,cos(\omega t) \tag{14.1}$$

Confined electrons with additional energy levels and changed of wave functions.

Take an electron (effective mass m_e^*) and a hole (effective mass m_h^*) interacting via Coulomb attraction screened by the static dielectric constant ε_r of the semiconductor. In SI units, the interaction potential is

$$V(\mathbf{r})=-\frac{e^2}{4\pi\varepsilon_0\varepsilon_r r} \tag{14.2}$$

where $\mathbf{r}=\mathbf{r}_e-\mathbf{r}_h$ is the electron–hole separation.

The two-body Hamiltonian (neglecting band-structure complications, exchange, and spin) is

$$\widehat{H}=-\frac{\hbar^2}{2m_e^*}\nabla_e^2-\frac{\hbar^2}{2m_h^*}\nabla_h^2-\frac{e^2}{4\pi\varepsilon_0\varepsilon_r\left|\mathrm{r}_e-\mathrm{r}_h\right|} \tag{14.3}$$

Next, we separate the center of mass and relative motion and define the center-of-mass coordinate **R** and relative coordinate **r**:

$$\mathbf{R}=\frac{m_e^*\mathbf{r}_e+m_h^*\mathbf{r}_h}{M},\ \mathbf{r}=\mathbf{r}_e-\mathbf{r}_h, M=m_e^*+m_h^* \tag{14.4}$$

The kinetic energy separates into a center-of-mass part and a relative part:

$$-\frac{\hbar^2}{2m_e^*}\nabla_e^2-\frac{\hbar^2}{2m_h^*}\nabla_h^2=-\frac{\hbar^2}{2M}\nabla_R^2-\frac{\hbar^2}{2\mu}\nabla_r^2 \tag{14.5}$$

where μ is the reduced mass:

$$\mu = \frac{m_e^* m_h^*}{m_e^* + m_h^*} \tag{14.6}$$

Thus, the Hamiltonian becomes

$$\widehat{H} = -\frac{\hbar^2}{2M}\nabla_R^2 + \widehat{H}_{\text{rel}}(\mathrm{r}),\ \widehat{H}_{\text{rel}}(\mathrm{r}) = -\frac{\hbar^2}{2\mu}\nabla_r^2 - \frac{e^2}{4\pi\varepsilon_0\varepsilon_r r} \tag{14.7}$$

The center-of-mass motion is free (plane waves). The bound exciton problem reduces to the relative-motion Hamiltonian $\widehat{H}_{\text{rel}}$, which is identical in form to the hydrogen atom Hamiltonian but with the electron mass replaced by the reduced mass μ and the Coulomb constant reduced by ε_r.

At the end, we acquire a hydrogen-like solution (energies and length scale).

The eigenvalues for the relative motion are the hydrogenic levels:

$$E_n^{(\text{rel})} = -\frac{\mu e^4}{2(4\pi\varepsilon_0\varepsilon_r)^2\hbar^2}\cdot\frac{1}{n^2}, n = 1, 2, \ldots \tag{14.8}$$

Define the *effective Rydberg* (binding energy scale).

$$\text{Ry}^* \equiv \frac{\mu e^4}{2(4\pi\varepsilon_0\varepsilon_r)^2\hbar^2} \tag{14.9}$$

Then,

$$E_n^{(\text{rel})} = -\frac{\text{Ry}^*}{n^2}.$$

The exciton Bohr radius (effective Bohr radius) is

$$a_B^* \equiv \frac{4\pi\varepsilon_0\varepsilon_r\hbar^2}{\mu e^2} \tag{14.10}$$

Equivalently, in terms of the hydrogen Bohr radius $a_0 = 4\pi\varepsilon_0\hbar^2/(m_0 e^2)$,

$$a_B^* = a_0\frac{\varepsilon_r m_0}{\mu}.$$

Often, it is convenient to express Ry^* in terms of a_B^*:

$$\text{Ry}^* = \frac{\hbar^2}{2\mu(a_B^*)^2} = \frac{e^2}{8\pi\varepsilon_0\varepsilon_r a_B^*} \tag{14.11}$$

We can now relate the solution to the ground-state exciton binding energy and radius.

For the ground state, $n = 1$:

- *Binding energy* (energy required to free the electron–hole pair)

$$E_b = \mathrm{Ry}^* = \frac{\mu e^4}{2(4\pi\varepsilon_0\varepsilon_r)^2\hbar^2} \tag{14.12}$$

This is usually quoted as a positive binding energy; the relative-motion eigenvalue is $-E_b$.

- *Exciton (Bohr) radius* (typical electron–hole separation)

$$a_X \equiv a_B^* = \frac{4\pi\varepsilon_0\varepsilon_r\hbar^2}{\mu e^2} \tag{14.13}$$

And further determine the exciton spectral position (relative to the band edge).

If E_g is the bandgap energy (energy difference between the conduction band minimum and the valence band maximum), the optical energy of the n-th exciton line (neglecting center-of-mass kinetic energy and small corrections) is approximately

$$E_X(n) = E_g - E_n^{(\mathrm{binding})} = E_g - \frac{\mathrm{Ry}^*}{n^2} \tag{14.14}$$

So the ground-state exciton optical line sits at $E_g - E_b$. One example is shown in Figure 14.1.

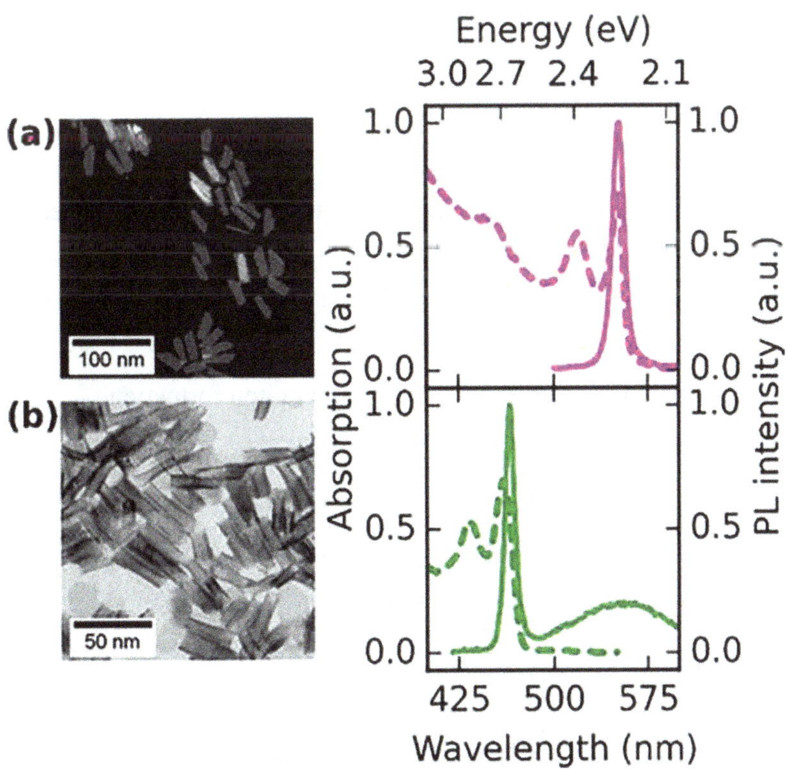

Figure 14.1 Different sizes of quantum wells demonstrating different exciton behavior.

ACS Nano 9 (12), 12473–12479 (2015). OA.

14.2 2D CRYSTALS AS NATURAL QUANTUM WELLS

A quantum well is a structure that confines carriers (electrons and holes) in one dimension while allowing them to move freely in the other two.

- Monolayer as a Quantum Well: It exists in stable form as a monolayer, which is only about thick. This atomic thickness is far smaller than the de Broglie wavelength of electrons and holes (typically several nanometers to tens of nanometers).
- Confinement: Because the thickness of the monolayer is on the order of the de Broglie wavelength, the motion of the carriers perpendicular to the plane (the z-direction) is quantized. The allowed energies for movement in this direction are discrete, leading to the formation of sub-bands instead of a continuous energy band structure found in bulk materials.
- Quantum Well Behavior: The carriers are thus confined within a potential well formed by the layer (the well material) sandwiched between the surrounding vacuum or substrate/capping layer (the barrier material).

The strong quantum confinement in 2D semiconductors leads to several enhanced properties, confirming their classification as quantum confined systems:

- Bandgap Transition: Bulk is an indirect bandgap semiconductor, but when reduced to a single monolayer, it transforms into a direct bandgap semiconductor. This change is a direct result of quantum confinement altering the relative energy levels at different points in the Brillouin zone.
- Exciton Binding Energy: The confinement significantly increases the electrostatic interaction between the excited electrons and the holes they leave behind. This results in extraordinarily high exciton binding energies (on the order of hundreds of meV), which means excitons remain stable and govern the optical properties even at room temperature, a behavior typical of strongly confined quantum structures.

In summary, the atomic-scale thickness of monolayers causes the charge carriers to be strongly confined in one dimension, leading to the quantization of energy levels and dramatic changes in electronic and optical properties. For these reasons, monolayers are appropriately categorized as two-dimensional quantum wells or, more broadly, two-dimensional electron gases (2DEGs).

Monolayer MoS_2 has a direct bandgap (~1.8 eV) at the K point of the Brillouin zone, which allows efficient exciton formation and radiative recombination. When MoS_2 absorbs a photon:

$$\text{Photon absorption} \rightarrow e^- + h^+ \rightarrow \text{Exciton formation.}$$

The exciton can decay into one of the following:

1. Radiatively: emitting a photon (PL)
2. Nonradiatively: energy lost to lattice vibrations (phonons) or defect states

The PL intensity I_{PL} is proportional to the number of radiative recombinations:

$$I_{\mathrm{PL}} \propto k_r n_{\mathrm{ex}},$$

where n_{ex} is the exciton population and k_r is the probability per unit time to recombine radiatively. Considering total recombination shown in Figure 14.2, the quantum yield (QY) can be expressed as follows:

$$\mathrm{QY} = \frac{I_{\mathrm{PL}}}{I_{\mathrm{abs}}} = \frac{k_r}{k_r + k_{nr}} \tag{14.15}$$

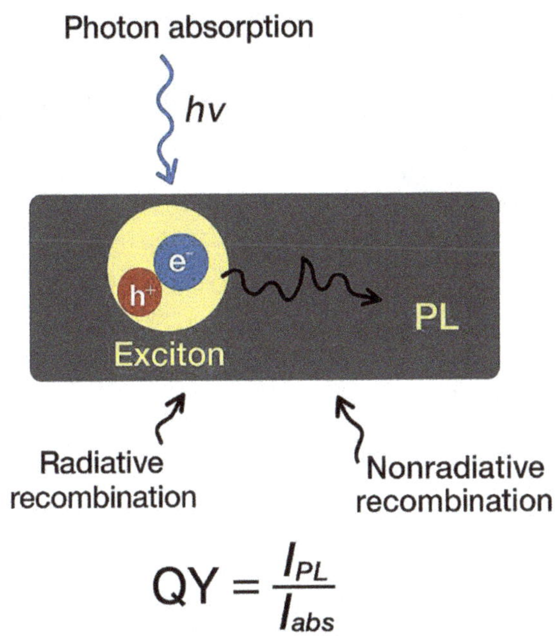

Figure 14.2 Recombination process for an exciton.

Diagram generated by AI assistance.

Particularly, a few factors can affect PL QY in MoS_2:

- Defects and traps: Sulfur vacancies and impurities introduce nonradiative pathways (k_{nr}), reducing QY.
- Substrate and environment: Interaction with the substrate or adsorbates can quench excitons. Encapsulation (e.g., in hBN) can increase QY by reducing nonradiative recombination.
- Exciton dynamics: Exciton–exciton annihilation at high excitation densities reduces QY. Trion formation (charged excitons) can shift emission and affect apparent QY.

14.3 QUANTUM DOTS

Bulk CdSe is a direct bandgap semiconductor with a bandgap of about 1.74 eV (~712 nm). When the particle size is reduced to the nanometer scale (typically 2–10 nm), comparable to or smaller than the exciton Bohr radius (~5.6 nm for CdSe), the motion of electrons and holes becomes spatially confined. This is called quantum confinement.

In this regime, the energy levels become discrete rather than continuous. The effective bandgap increases as the particle size decreases, and the energy of the first allowed electronic transition is approximately

$$E_g(R) = E_g^{\text{bulk}} + \frac{\hbar^2\pi^2}{2R^2}\left(\frac{1}{m_e^*} + \frac{1}{m_h^*}\right) - 1.8\frac{e^2}{4\pi\varepsilon_0\varepsilon R} \tag{14.16}$$

where R is the quantum dot radius, m_e^* and m_h^* are the effective masses of the electron and hole, ε is the dielectric constant of CdSe, and the first term is the quantum confinement energy and the second term is the Coulomb interaction energy.

As R decreases, $E_g(R)$ increases, resulting in a blue shift in absorption and emission. For their absorption properties:

- **Size-dependent absorption:** Smaller CdSe QDs absorb higher-energy (shorter wavelength) light due to the larger effective bandgap.
- **Discrete excitonic peaks:** At low temperatures and for monodisperse samples, the absorption spectrum shows sharp peaks corresponding to discrete exciton transitions.
- **Extinction coefficient:** CdSe QDs have large molar extinction coefficients ($\sim 10^4$–10^5 M^{-1} cm^{-1}), making them highly absorbing.

CdSe QDs exhibit *strong photoluminescence*, and the emission wavelength can be tuned from *~450 nm (blue) to ~650 nm (red)* simply by changing the particle size, as shown in the growth method in Figure 14.3. Key points include the following:

- Size-tunable emission: Small QDs → higher bandgap → blue emission; large QDs → lower bandgap → red emission.
- QY: PL efficiency depends on surface passivation and defects. QDs with ligand passivation or core-shell structures (e.g., CdSe/ZnS) can achieve QY > 50–90%.
- *Narrow emission peaks*: Typical full width at half maximum (FWHM) ~20–30 nm, ideal for color display applications.

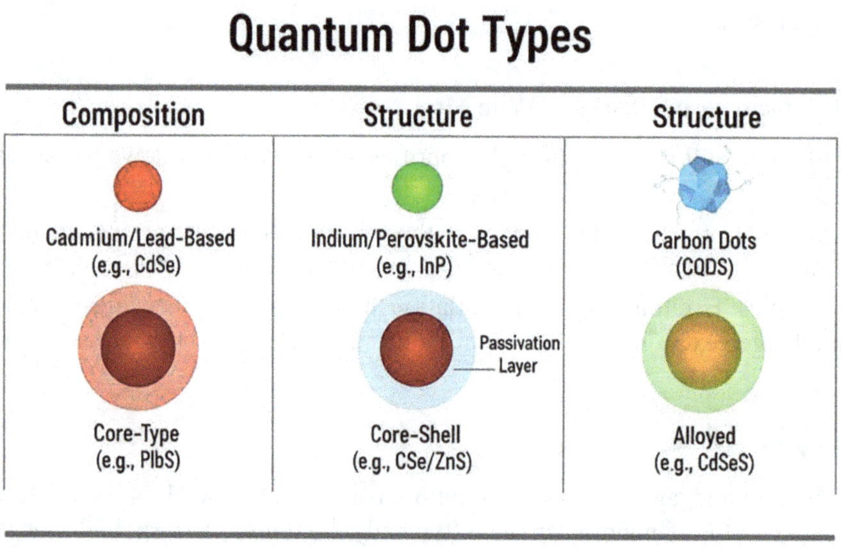

Figure 14.3 Bottom-up method capable of tuning the sizes of quantum dots.

Diagram generated by AI assistance.

CHAPTER 15

Supplemental Figures

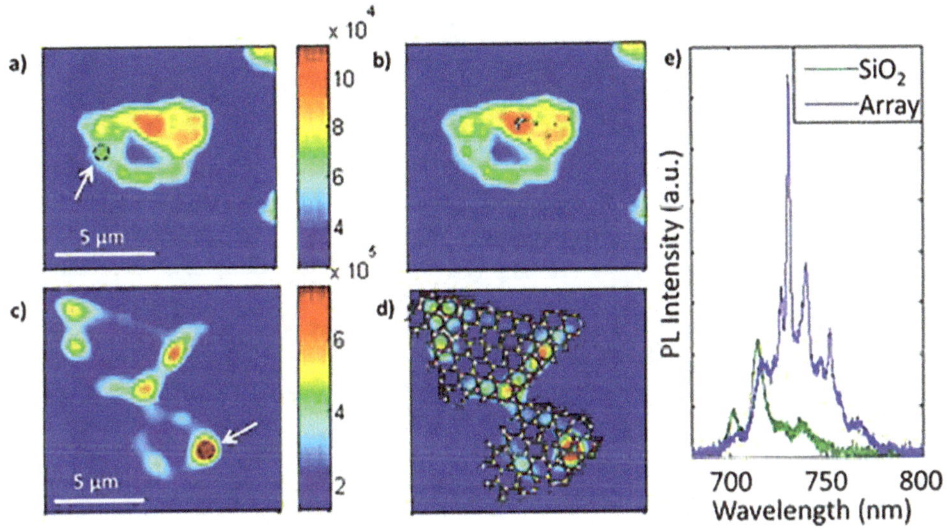

Figure 15.1 PL change between Si and silver arrays.

Nano Lett.17 (7), 4317–4322 (2017).

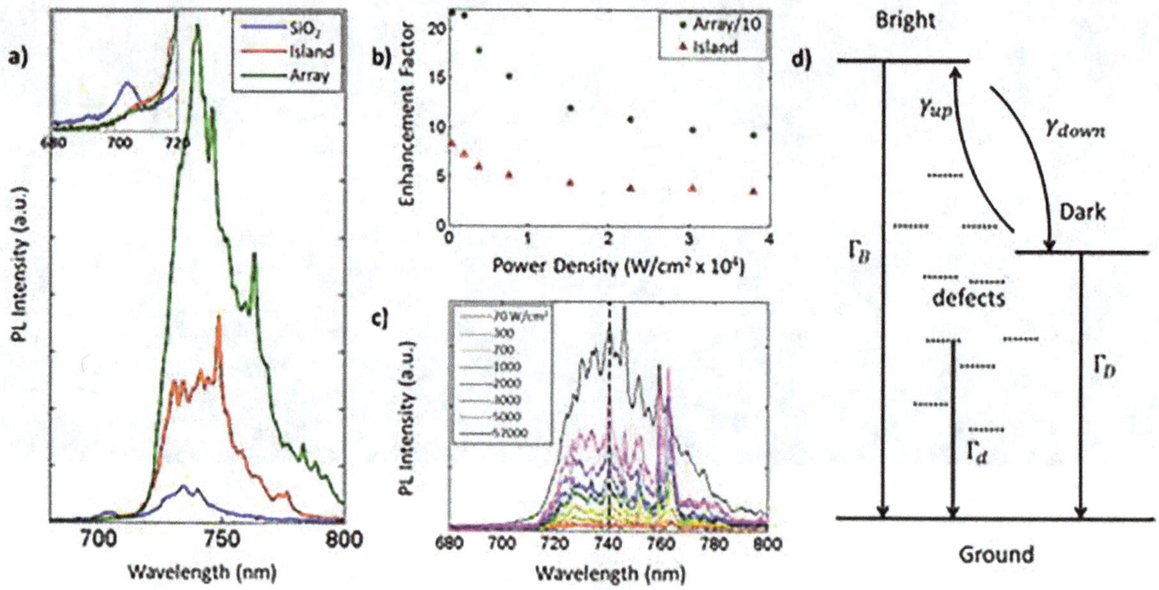

Figure 15.2 Power-dependent PL of WSe2.

Nano Lett.17 (7), 4317–4322 (2017).

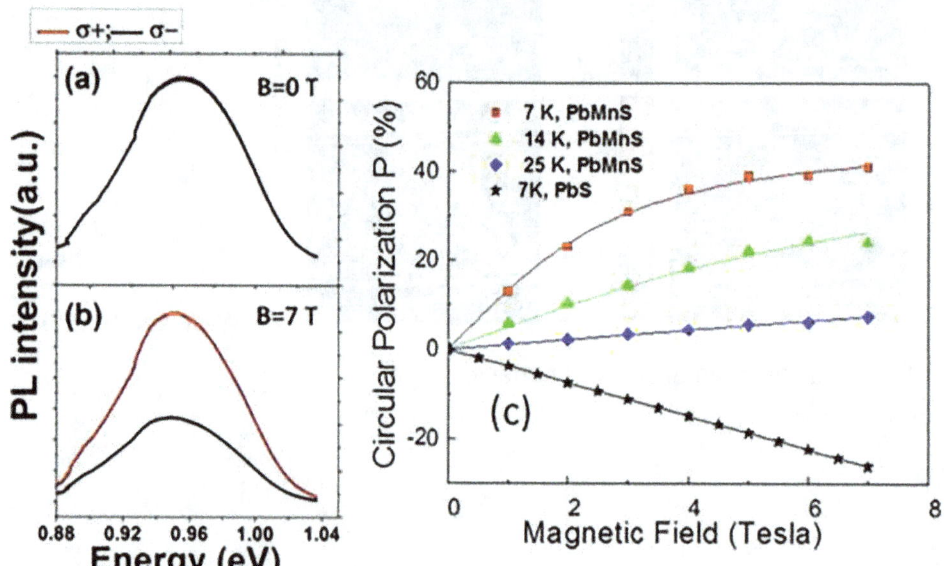

Figure 15.3 PL circular polarization changes from magnetic quantum dots.

Appl. Phys. Lett. 101, 062410 (2012).

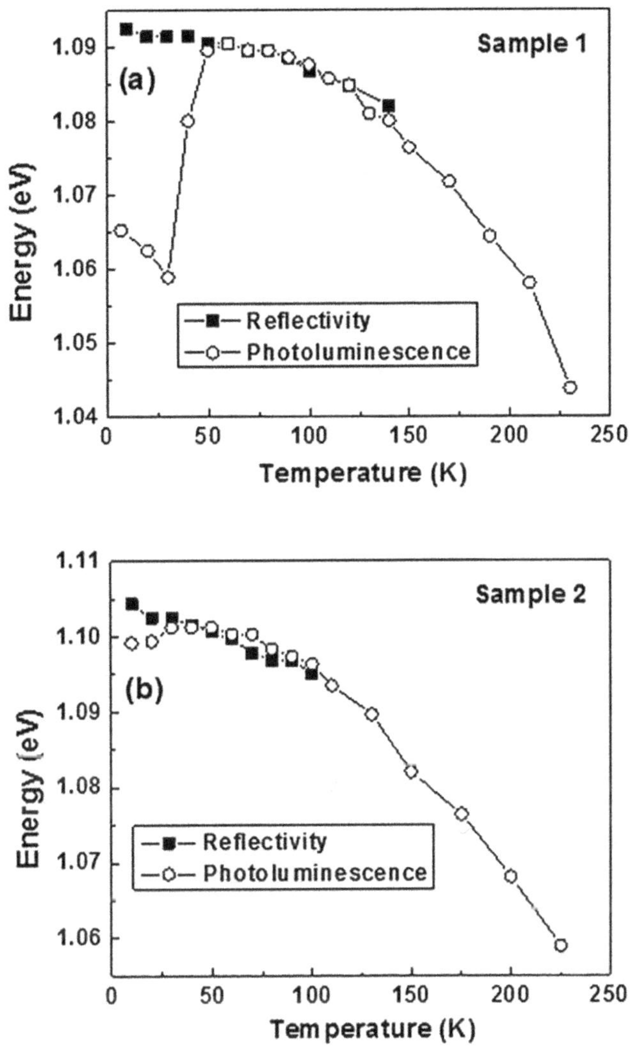

Figure 15.4 Temperature dependence of PL vs. reflectivity in GaAs InN layer.

Appl. Phys. Lett. 103, 012104 (2013).

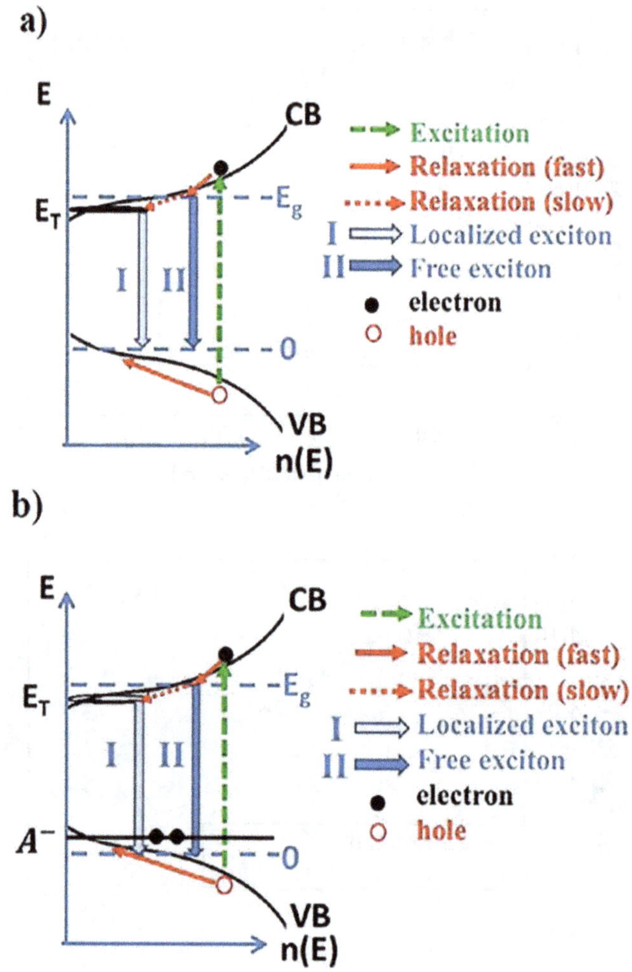

Figure 15.5 Possible energy band discussion based on optical data in Figure 15.5.

Appl. Phys. Lett. 103, 012104 (2013).

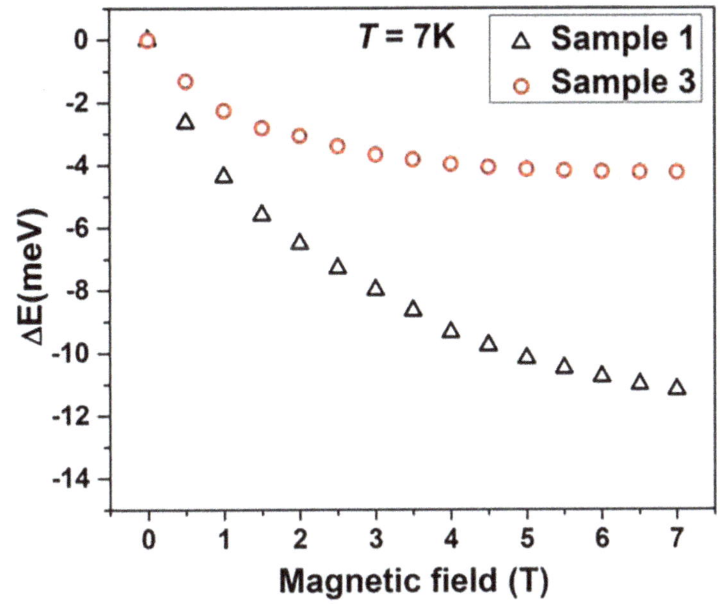

Figure 15.6 PL peak shift with applied magnetic fields.

Phys. Rev. B 92, 035430. OA (2015).

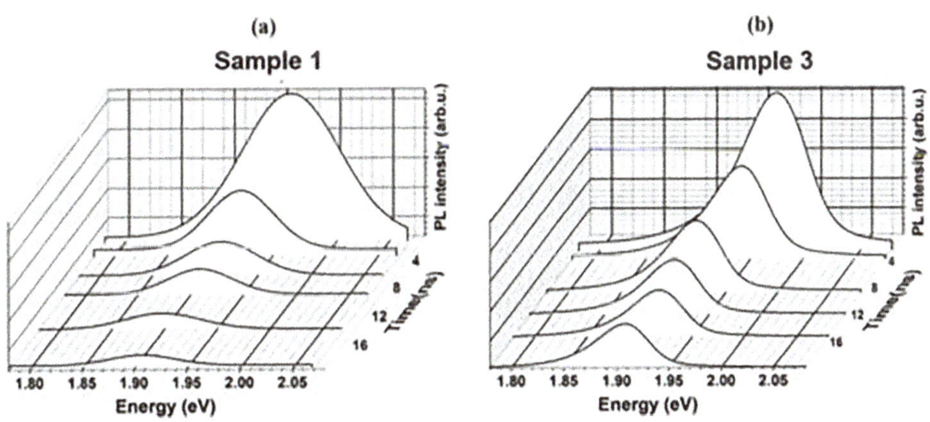

Figure 15.7 PL decay behavior.

Phys. Rev. B 92, 035430. OA (2015).

www.ingramcontent.com/pod-product-compliance
Lightning Source LLC
LaVergne TN
LVHW082354200926
848481LV00012B/3557
9798319712226